*Introduction
to the*
# Theory of Constraints (TOC) Management System

# The St. Lucie Press/APICS Series on Constraints Management

## Series Advisors

**Dr. James F. Cox, III**
*University of Georgia*
*Athens, Georgia*

**Thomas B. McMullen, Jr.**
*McMullen Associates*
*Weston, Massachusetts*

## Titles in the Series

**Introduction to the Theory of Constraints (TOC) Management System**
*by Thomas B. McMullen, Jr.*

**Securing the Future: Strategies for Exponential Growth Using the Theory of Constraints**
*by Gerald I. Kendall*

**Project Management in the Fast Lane: Applying the Theory of Constraints**
*by Robert C. Newbold*

**The Constraints Management Handbook**
*by James F. Cox, III and Michael S. Spencer*

**Thinking for a Change: Putting the TOC Thinking Processes to Use**
*by Lisa J. Scheinkopf*

# Introduction to the
# Theory of Constraints (TOC) Management System

Thomas B. McMullen, Jr.

The St. Lucie Press/APICS Series on Constraints Management

APICS®
The Educational Society for Resource Management

S$_L^t$

St. Lucie Press
Boca Raton   Boston   London   New York   Washington, D.C.

Library of Congress Cataloging-in-Publication Data

McMullen, Thomas B.
    Introduction to the theory of constraints (TOC) management system
  /  by Thomas B. McMullen, Jr.
       p.  cm. — (St. Lucie Press/APICS series on constraints management)
    Includes bibliographical references and index.
    ISBN 1-57444-066-7
    1. Theory of constraints (Management)  I. Title.  II. Series.
  HD69.T46M38  1998
  658.5—dc21                                     98-10293
                                                          CIP

"Different People" words and music by Gwen Stefani and Tony Kanal. © Copyright 1995 by MCA Music Publishing, a Division of Universal Studios, Inc. (ASCAP)/Knock Yourself Out Music (ASCAP). International copyright secured. All rights reserved. Used by permission.

"World Go Round" words and music by Eric Stefani, Gwen Stefani, and Tony Kanal. © Copyright 1995 by MCA Music Publishing, a Division of Universal Studios, Inc. (ASCAP)/Knock Yourself Out Music (ASCAP). International copyright secured. All rights reserved. Used by permission.

Sarah McLachlan's "Wait" copyright © 1993 Sony/ATV Songs LLC/Tyde Music. All rights administered by Sony/ATV Music Publishing, 8 Music Square West, Nashville, TN 36203. All rights reserved. Used by permission.

# Dedication

To Carol Charpie McMullen, a beautiful woman,
wife, mother, partner, and friend:
I may have mentioned once, Carol, that
it's not easy for a fella to count stars alone.
Thanks for helping me get it done.


And to the memory of a beautiful
woman, mother, mentor, and
friend — Barbara Yvonne McMullen:
Y'know, Mom, if we'd started our little project
fifteen years from now, or even today,
instead of fifteen years ago, we'd have gotten it done.
(It wasn't inherently that tough a problem.)

# Contents

List of Quotations ........................................................................................ xviii
List of Figures ............................................................................................... xix
About APICS ................................................................................................ xxi
The Author .................................................................................................. xxii
Acknowledgments ........................................................................................ xxiv

Introduction: A Physics of Anything ............................................................ 1
  Book's Purposes .......................................................................................... 1
  What's All the Excitement About? ............................................................... 2
    New Management Science ...................................................................... 2
    TOC's Version of the Scientific Method .................................................. 2
    These Ivory Tower Debates Just Don't Matter in the Field ..................... 3
  Good Problem To Have? ............................................................................. 3
    What's Gotten into These People?!!! ...................................................... 3
      They're Suddenly Demanding Better Solutions ................................... 3
      But They're Also Delivering Better Solutions ...................................... 4
      Perception of Possibilities ................................................................... 4
      Confidence and "Do It Now!" Culture Restored .................................. 4
      Old Guard and New Guard: New and Powerful Synthesis .................... 5
      Is it any Wonder? .............................................................................. 6
    Problem or Opportunity? ...................................................................... 6
      Problem? .......................................................................................... 6
      Opportunity? ..................................................................................... 6
      This Book Helps Executives and Directors Sort it Out ........................ 6
  What Is Your Organization's Foundation? ................................................... 7
    Management System's Foundations ........................................................ 7
    Infrastructure Supporting Growth ......................................................... 7
    Global Standards and Timing ................................................................ 7
    "Knowledge Worker" Productivity ........................................................ 7

Integrated System vs. Happenstance Sub-Systems ............................................ 8
The Mind of the Strategist .................................................................................. 8
Blasting the Constraints on Using TOC ................................................................. 8
Who Should Read This Book? .............................................................................. 8
The Thoughtful and Practical "Managers" in Life ......................................... 8
Directors and Senior Managers Included ..................................................... 9
Breaking TOC's "Due Diligence" Constraint .................................................. 9
The "Airplane Book" ..................................................................................... 10
Let's Get This Show on the Road! ........................................................................ 10
Let's Use TOC in the Workplace ........................................................................ 10
"Saving the World" is Back in Style ................................................................... 10
Enough Already with the "Powerlessness" Story! ............................................. 11
Making the Difference We Can Make: It's Bigger Now ................................... 12
Let's Get to Work! ............................................................................................... 13
Keep Me Posted .................................................................................................. 13

1 What Is the Theory of Constraints? ...................................................... 15
A Management Science ........................................................................................ 15
A Physics of Anything ........................................................................................ 15
... A Walk Through a Business Bookstore, Circa 1997 .............................. 16
While We're At It Here, What's Wrong With This Picture? ...................... 19
... Physics of Performance Improvement ..................................................... 20
Physical and Thinking Constraints .................................................................... 20
Scope of Physics Widened ................................................................................. 21
Validity vs. Truth: A Core TOC Concept and Criterion ................................. 21
Another Core Concept: Verbalizing Intuition .................................................. 21
Everyone a Physicist .......................................................................................... 22
Beyond Correlation to Cause-and-Effect .......................................................... 22
Generalists Provide "Guidance" to Specialists ................................................. 22
Win-Win Solutions ............................................................................................. 22
Making Common Sense a Common Practice ..................................................... 23
Starting Right Is Half the Battle ........................................................................ 23
TOC Management System .................................................................................. 23
TVA Financial Management System ................................................................. 24
Management System of Choice .......................................................................... 25
There Is a Best Way To Begin To Think About TOC ...................................... 25
From Overview to Elements ............................................................................... 25
Elements of the TOC Management System ......................................................... 26
Diving Into Our First TOC "Logic-Tree" Diagram ........................................ 26
TOC Systems Thinking Tools ............................................................................ 26
Verbalizing Intuition Leads to Systems Thinking ....................................... 28
Intrinsic Order Principle Also Leads to System Concept ........................... 28
A Typical TOC Systems Thinking Project ..................................................... 28

Defining the System ................................................................................ 29
Goal of System ...................................................................................... 29
Storyline of the Primary TOC Book, *The Goal* .................................... 30
Goal = Make Money .............................................................................. 30
Important Class of Organizations ......................................................... 31
Other Systems, Other Goals ................................................................. 32
    Different System Types, Goals, and Measurements ....................... 32
System's Constraints ............................................................................. 32
Measurements of the Goal .................................................................... 33
    TOC's Three Fundamental Measurements ..................................... 34
    If Money Is Not the Goal, Then Make Money the Operational Goal ....... 35
Scales of Importance, Part I (Global vs. Local Measurements) ............ 35
Scale of Importance, Part II (TVA Throughput World) ....................... 36
Necessary Conditions ........................................................................... 36
Combining Financial and Non-Financial Measurements ..................... 37
The Ultimate Process Improvement Project: Improving
the Process of Ongoing Improvement Itself .......................................... 37
    Three Questions and Roles for Management ................................. 37
    Focusing the Efforts ...................................................................... 38
    Improvement Curves ..................................................................... 39
    Apply Improvement Curves to Anything ...................................... 40
        Sample Improvement Curve: Little League Baseball Batting ....... 41
    Time Is the Prime Constraint ........................................................ 41
    Inherently Best Improvement Process ........................................... 42
    TOC's Five-Step Focusing Process ................................................ 42
    Using the Five Steps ...................................................................... 43
    General Purpose Process ................................................................ 43
    Three (of Five) Steps To Begin Lean and Agile Manufacturing ....... 43
    Uncovering Hidden Capacity and Process Improvement
    in Manufacturing ........................................................................... 44
    New Standards for Using Time Well .............................................. 45
Intrinsic Order Principle ....................................................................... 45
    Logic-Tree Thinking Processes
    Reveal Intrinsic Order ................................................................... 46
TOC Thinking Processes (Logic Trees) ................................................. 47
*Mission Impossible:* The Movie ............................................................ 47
    There's Thinking and Then There's *Thinking* ............................... 48
*Mission Impossible:* The TOC Case Study ............................................ 49
    Tool Selection Time ....................................................................... 49
Summary of TOC Logic-Tree Thinking Processes ................................ 50
Benefits of the Logic-Tree Thinking Processes ..................................... 51
    Let's Pay a Visit to Ethan Hunt ..................................................... 51
Four Stages of Struggle, Practice, and Skill with Logic Trees ............. 52

Stage One ............................................................................................... 52
Stage Two .............................................................................................. 52
Stage Three ........................................................................................... 53
Stage Four ............................................................................................. 54
Ethan Hunt Uses TOC Logic Trees? ................................................ 55
Current-Reality Tree Process and Diagram ............................................ 55
The Heart of the Matter ...................................................................... 55
The Facts Line Up at Harvard Business School ............................. 58
Several Ways To Begin Current Trees ............................................. 58
Building a TOC Current-Reality Tree ............................................... 59
In Better Shape Now ............................................................................ 60
Future-Reality Tree (FRT) Process and Diagram ................................. 61
Sample TOC Future Tree: TOC Management System ......................... 62
Generic vs. Situation-Specific Logic Trees ..................................... 63
Logic-Tree Format ................................................................................ 63
Level of Aggregation of Logic Trees ................................................ 63
Viewpoint of Logic Trees .................................................................... 63
The Example ........................................................................................... 64
Ethan's First Future-Reality Tree (Plan) Treats Only Symptoms .............. 65
Evaporating-Cloud (Conflict) Logic-Tree Process and Diagram .................. 65
Fresh Out of Ideas? Break Out the Evaporating-Cloud Process .............. 65
Example of Needs-and-Wants Cloud Format ................................... 66
Other Formats for Evaporating-Cloud Logic Trees ...................... 69
Ethan Uses the Objective-Requirements-Prerequisites Format ................ 69
Injections from Cloud Process Form New Future Tree .............................. 70
Could Have Gotten There More Quickly ......................................... 70
Failed to Apply TOC's CLR ................................................................ 71
Categories of Legitimate Reservations ........................................... 71
Encourage Study of the TOC Logic-Tree Rules ........................... 72
But, For Heaven's Sake, Don't Tie the Do-ers Down! ................. 73
Learn True TOC Logic-Tree Process From the Do-ers .............................. 74
Ethan Returns to and Corrects the Current-Reality Tree ......................... 74
Prerequisites Tree (PRT) Process and Diagram .................................... 74
Touring the Sample Prerequisites Logic Tree ............................... 79
Transition Logic-Tree Process and Diagram ........................................ 80
Logic-Tree Tools for Successive Levels of Detail ........................ 82
One Intermediate Objective Is Tricky and Important Enough
To Warrant the Additional Detail of a Transition Tree ............................. 82
Better Use of Meetings ............................................................................... 84
TOC Thinking Processes: Time Is the Prime Constraint ......................... 84
Initial Frustration, but Worth the Trouble ..................................... 85
Applications of TOC Logic-Tree Processes .......................................... 85
Other Thinking Processes ........................................................................... 85

    Ben Franklin's Decision Process ...................................................... 86
    Lastname1-Lastname2 .................................................................... 86
    Total Quality Management ............................................................. 87
    Systems Dynamics and *Fifth Discipline* ....................................... 87
    TRIZ ............................................................................................. 87
TOC Thinking Processes and Day-to-Day Management Skills ........... 88
TOC Project Management ...................................................................... 89
  Overview of TOC and Project Management ...................................... 89
    General ......................................................................................... 89
    Strategy ......................................................................................... 89
    Process Improvement .................................................................... 89
    Range of Applicability .................................................................. 89
    Critical Chain ............................................................................... 89
  Keeping Projects on Track ............................................................... 90
    It's Awful When a Project Starts Wrong or Dies Too Late .......... 90
    TOC Logic Trees .......................................................................... 91
    Airlines Computer System ............................................................ 92
  The Most Fundamental TOC Projects: Making Strategic Moves .... 92
    Some Projects Are More Important Than Others ......................... 92
    TOC Strategy Process Tools Help with Any and All .................... 93
    Understanding What You Are Seeing in Industry and Competition ........ 93
    Scenarios: *The Art of the Long View* ........................................... 93
    Avoiding Layoffs .......................................................................... 94
    Other Considerations .................................................................... 94
  Process Improvement Projects ......................................................... 95
    The Search for Low-Hanging Fruit ............................................... 95
    Pick a Core Process ...................................................................... 96
    The Right Way .............................................................................. 96
  Critical Chain vs. Critical Path ........................................................ 96
TVA Financial Management System ...................................................... 97
  The Need To Create and Measure Value ......................................... 97
  TOC Financial Foundations ............................................................. 97
  TOC Throughput Value Added (TVA) ............................................. 97
  Making the Pie Bigger ..................................................................... 99
  Scale-of-Importance Issues .............................................................. 100
  Better and Easier Than Allocation-Based Approaches .................... 102
  TVA's Advantages Over EVA ........................................................... 102
  Inviting Detective Columbo To Help Build Consensus? .................. 103
  Need To Know How Much You Can Make ....................................... 103
TOC Logistics and Factory Scheduling .................................................. 104
  Overview of TOC Industrial Theory and Terminology .................... 104
    Factory Machines with Infinite Capacity? .................................... 104
    Factory Machines with Finite Capacity ........................................ 105

Every Factory has a Drum, a Herbie ............................................................ 105
Drum-Buffer-Rope (DBR) Scheduling ........................................................ 106
Synchronous Manufacturing ...................................................................... 106
Buffer Management Shopfloor Control ...................................................... 106
Early Warning System................................................................................ 107
Factory Makes More Money ...................................................................... 107
Third Capacity Category: Protective Capacity .......................................... 107
Time Buffers.............................................................................................. 108
Dynamic Buffering .................................................................................... 108
Finite Capacity Replaces Infinite Capacity Processes .............................. 108
Maximum Manufacturing Company Return-on-Investment ................. 109
TOC Applied to the Manufacturing, Enterprise, Distribution,
and Supply Chain .................................................................................. 109
That's It ...................................................................................................... 109
This Has Been a Goldratt Baseline TOC (GBT) ............................................ 110
Summary ........................................................................................................ 110
The Theory of Constraints Management System:
Emerging Global Standard ........................................................................ 110
Hits the Bull's-Eye for Managing in the Twenty-First Century.................... 111

2  Where Did TOC Come From? ........................................................................ 113
Answer One: Eliyahu M. Goldratt, Ph.D. .................................................... 113
The Early Years .......................................................................................... 113
Breakthroughs in Scheduling .................................................................... 114
Creative Output, Inc. ................................................................................ 114
A Business Novel, a Love Story, and a Surprise ........................................ 115
Leaves One Company ................................................................................ 115
Starts Another Company............................................................................ 115
Solves Cost Accounting Problem .............................................................. 116
Teaching the World How To Think? .......................................................... 116
TOC Logic-Tree Thinking Processes ........................................................ 116
Logic Trees Are New Expression of Scientific Method .............................. 117
Project Management .................................................................................. 117
What Is After "Retirement"?...................................................................... 117
Answer Number Two: Intrinsic Order — TOC Was Always
Right There Waiting .................................................................................... 117
Answer Number Three: Lots of People Discovered TOC, or Parts of TOC ...... 118
Answer Number Four: Many People Worked
With Dr. Goldratt on Developing TOC ........................................................ 119
Answer Number Five: "Graduate Students" Continue Independently .............. 120
Bottom Line: Dr. Goldratt Has Provided the Leadership .................................. 121
Sharing the Process of Genius........................................................................ 122
As a Physicist.............................................................................................. 122

As an Inspirational Motivator ................................................................ 122
As a Philosopher ................................................................................... 122
As a Practical Kind of Guy ................................................................... 122
As a Self-Directed Change Agent ......................................................... 123
As a Genius .......................................................................................... 123

**3  Why Do People Say They Like TOC?** ................................................... 125
TOC Management System Makes Common Sense a Common Practice ......... 125
TOC Increases Profitability ................................................................... 125
TOC Exposes Hidden Capacity ............................................................. 126
TOC Makes Growth (in TVA) the Perpetual Top Priority ..................... 127
TOC Improves the Profitability of Product Mix .................................... 128
TOC Focuses Manufacturing and Quality Improvements on TVA .......... 128
TOC Reduces Inventories in Supply Chain ........................................... 129
TOC Establishes Excellent Delivery Performance .................................. 130
TOC Produces Better Returns-on-Investment (ROI) ............................. 131
TOC Saves Factories and Jobs .............................................................. 131
TOC Saves Yet Another Factory ....................................................... 132
TOC Thinking Processes ............................................................... 132
The Drums of Drum-Buffer-Rope ................................................. 132
Buffer ......................................................................................... 132
Rope ........................................................................................... 133
Synchronous Manufacturing ........................................................ 133
Results ........................................................................................ 133
But It Could Have Been Different ................................................. 133
Another Way It Could Have Been Different .................................. 134
TOC Reduces and, Ideally, Eliminates the Need for Layoffs ............. 134
Growth in Cash Flow (TVA) — A Perpetual High Priority ............... 137
TOC Makes Growth the Perpetual First Priority .......................... 137
Calling All Trade Unions: Tell Managements You Think TOC
Is a Good Way To Go .................................................................. 137
TOC Creates Better Workplaces ........................................................ 138
What's at Stake? .............................................................................. 138
TOC Makes Common Sense a Common Practice ............................. 138
Things "Make Sense" When Coherence Exists ................................. 139
Constructing "Common Sense" in Situations .................................. 139
Increased Ability To Test New Ideas ............................................... 139
TOC Focuses the Ideas of Systems Thinking .................................. 139
TOC Improves Communication and Build Teams ............................ 140
TOC Allows Verbalizing Intuition To Play an Appropriate Role ........ 140
Focusing All Those Positive Attitudes ............................................. 141
TOC Guides and Focuses the Idealist and Empowerment Approaches ...... 141
Overlaying Focus Onto the Chaos? ............................................. 141

TOC Makes Idealist Approaches Safe and Effective .................................. 142
Rebalances Power Among Types ....................................................... 142
TOC Empowers People To Create Their Own Best Practices ...................... 143
TOC Encourages People To Invent New Best Practices ........................... 143
TOC Encourages People To Propose Solutions ................................... 144
Better Solutions .......................................................... 144
TOC Creates True Learning Organizations .................................. 144
Field Research in Industrial Settings ...................................... 144
In Short: It Works! ...................................................... 144

**4  Who Has Used TOC?** ..................................................... 147
By Now, Too Many To Tell of All Its Uses ................................... 147
Rationale for Representative Sample ....................................... 147
APICS Constraints Management Symposia .................................... 148
1995 APICS Constraints Management Symposium ....................... 149
1996 APICS Constraints Management Symposium ....................... 151
1997 APICS Constraints Management Symposium ....................... 156
Additional Sources ....................................................... 159
APICS International Conferences ...................................... 159
Institute for International Research ................................... 159
Institute of Management Accountants Report ........................... 161
Theory of Constraints Field Study Report ......................... 161
IMA Global Solutions Conference Series ................................ 162
International Society for Systems Improvement ........................ 162
Clemson University .................................................. 163
Other Public Domain Indications of TOC Applications ............... 163
Summary ................................................................. 168

**5  Where Is TOC Headed?** ................................................ 171
Any Important and Complicated Problem Will Do ........................... 171
Bigger and Tougher Problems Can Reasonably Be Tackled ............... 171
"The Wisdom To Know the Difference" ................................. 171
On the One Hand, There Is Nothing New ............................... 172
On the Other Hand, Something Is New ................................. 172
More Skills Mean More Is Possible ................................... 172
The Wisdom To Know the Difference: Revisited ...................... 173
Best of Times or Worst of Times? .................................... 173
The World Is Smaller Now ................................................ 174
TOC Thinking Processes to the Rescue! .................................. 175
Activists Will Like the Thinking Processes ......................... 176
Conservatives Will Like the Thinking Processes ..................... 176
Obvious Fertile Fields for TOC Work .................................... 176
Thinking and Health .................................................... 176

    Dr. Herbert Benson: Applying Science to Non-Mainstream Methods ........ 177
    Simple Communication .................................................................... 177
    Mental Health Support and Advocacy ..................................... 178
  Ecumenism ............................................................................................. 178
  Environment .......................................................................................... 179
    Permanent Condition: No Turning Back ................................. 179
TOC and Information Systems ............................................................... 179
  World Wide Web Becomes Worldwide Forest — of TOC Trees! ................. 179
  TOC Project Management Systems ................................................. 180
  TOC Manufacturing Systems: *De Facto* Co-Standard .................... 180
  Beyond TOC Manufacturing Information Systems ..................... 183
TOC for Education ................................................................................... 183
  Middle School Children Show Industry How It Is Done ............... 183
  TOC for Education, Inc. ..................................................................... 184
    High School's Peer Mediation Program ................................... 185
    Clouds for Everywhere? .................................................................. 186
    Preventing Conflicts ..................................................................... 186
    Teacher's Role: To Teach Students How To Think .................. 186
    Using TOC To Design — and Within — Curricula ............... 186
    Cause-and-Effect Thinking Delivers Better Student Essays ..... 187
    Establishing Respect for Teacher in East L.A. Classroom ......... 187
    Elementary School Principal Says TOC Is a Foundation for Change ..... 187
    Students Plan Together .................................................................. 188
    Impact Within Juvenile Detention Camp .............................. 188
  Harnessing Emotion and Putting It to Constructive Use ............ 188
  Avery-Dennison Manager Takes Aim at University Education .......... 189
  Harvard Business School: Logic Trees for the Case Method? ......... 189
  U.S. Naval Academy: Logic Trees for Policy and Strategy? ............ 189
  People Who Rely on Formal Education Will Fall Behind ............ 190

**6 How Do I Get Started With TOC?** .................................................... 193
Five Steps To Introduce the TOC Management System ..................... 193
  Preliminary Words of Advice ........................................................... 194
    Lessons of Experience .................................................................. 195
    Objectives of the Implementation Process ............................ 195
    How Much TOC, at What Level of Detail, Is Enough? ........... 195
    TOC Educators and Consultants ............................................. 197
Step One: TOC Education ..................................................................... 197
  Use the Primary TOC Business Novels Well .................................. 197
  Build the Company TOC Resource Library .................................... 199
    APICS TOC Conference Symposium Proceedings and Audiotapes ........ 199
    TOC Publications ......................................................................... 199
    APICS TOC Resources Available via APICS Online ................. 204

TOC on the Internet ..................................................................................... 204
Conduct TOC Education and Training ......................................................... 204
Join APICS and the APICS Constraints Management SIG ...................... 205
Use APICS Constraints Management (TOC) Symposia .......................... 205
Use the TOC Portion of the APICS International Conference ................ 205
Use the APICS Theory of Constraints Workshops ................................. 205
Make Use of the APICS Theory of Constraints Certifications................. 206
Use the TOC Logistics Games ................................................................ 206
The Dice Game ....................................................................................... 206
Computer-Supported TOC Simulations ................................................. 207
Drum-Buffer-Rope (DBR) Game Series ................................................. 207
Setup for the Drum-Buffer-Rope Game ........................................... 207
The Action ...................................................................................... 208
Step Two: Integrate TOC With Other Management Approaches ..................... 208
Use TOC in Support of Covey's Seven Habits ............................................. 209
Use TOC To Improve a Learning Organization's Foundation and Process ...... 212
Use Logic Trees To Capture Individual and Organizational Learning......... 212
Use TOC To Guide Total Quality Processes ..................................... 213
Use TOC To Coordinate the Competing Concepts
in Manufacturing Management ................................................................ 213
Just-in-Time ........................................................................................... 214
Shingo Management Methods ................................................................ 215
Support Your MRP II and ERP Systems with TOC's Solution
to the Famous "Infinite Capacity" and "Work Order" Problems ............ 215
Use TOC To Guide and Shape Meaningful Coincidence .............................. 216
Some General Advice: Other Management and Improvement Approaches .... 216
Step Three: Build a TVA Financial Management System ................................... 217
Invite Detective Columbo in To Introduce TOC Accounting ...................... 217
Why Does TOC Not Treat Factory Direct Labor as a Variable Cost?....... 224
Next Step: Build the TVA Financial System .................................................. 225
Don't Make Management Accounting Difficult: Use TVA ...................... 226
Step Four: Use TOC for Projects Other than Major Changes
to the Logistics Infrastructure ......................................................................... 226
Keep Track of TOC Interest and Use .......................................................... 226
Encourage the Executive Management Team To Start Using
the TOC Logic Trees .................................................................................... 226
Every Manager Develops Trees ............................................................. 228
Role Models for Executive Use of TOC Thinking Processes ................... 228
Early Executive Use of TOC Logic Trees To Maintain and Increase
TVA Avoids Punishing People for Doing the Right Things........................... 229
Use TOC Logic Trees and TVA-I-OE Measures for Strategy ....................... 230
Use TOC To Improve Competitive Monitoring............................................. 230
Commercial, Legal, Diplomatic, and Military Applications.................... 230

More Effective Use of Brilliant Strategists ............................................. 231
   Mutual Fund Industry ......................................................... 231
   The Scenario ..................................................................... 231
Use TOC Best Practices in Strategy ............................................... 232
Use TOC To Improve a Process ..................................................... 234
   Use TOC and TP as the Process To Select Processes
   for Process Management, Improvement, or Reengineering ..................... 234
   Effective Business Process Design ............................................. 235
Use TOC Thinking Processes in Place of Benchmarking and Surveys ......... 235
Encourage Use of TP in Management Skills Format ................................... 236
Step Five: The Major TOC Logistics Projects ...................................... 236
   Build Drum-Buffer-Rope and Other Useful Logistics Systems on Your
   Company's New Foundation of TOC Processes ............................... 236
   Establish TOC Information Systems To Support
   the New Logistics Infrastructures ............................................. 236
When the Five Implementation Steps Are Complete ........................... 236
   What If My Company Is Not Introducing TOC? ............................. 237
Seven Things You Can Do Right Away! ............................................. 237
   Do These Seven Things Right Away! ......................................... 237
   Use This Book To Get Started Right Away .................................. 240
   Trees by Hand or Through the Use of Systems? Yes. ...................... 241
No Time Like the Present! ............................................................... 241

Epilogue ......................................................................................... 243
A Riddle, But Follow the Rules ....................................................... 243
Thank You for Taking the Time ....................................................... 243

Notes ............................................................................................. 245
Notes for the Introduction ............................................................. 245
Notes for Chapter 1 ..................................................................... 252
Notes for Chapter 2 ..................................................................... 263
Notes for Chapter 3 ..................................................................... 264
Notes for Chapter 4 ..................................................................... 276
Notes for Chapter 5 ..................................................................... 277
Notes for Chapter 6 ..................................................................... 278

Index ............................................................................................. 281

# List of Quotations

Hannibal Smith (Plan Comes Together) ............................................................ xxvi
Ralph Waldo Emerson (Success) ...................................................................... xxvi
Stefani, Stefani, and Kanal (Those Little Things Matter) ................................. xxvi
George Bernard Shaw (Thinking and Thinking) ................................................ 14
Plato (A Pilot's Art) ......................................................................................... 14
Confucius (Original Logic Trees?) ................................................................... 14
Dr. Eliyahu M. Goldratt (All Be Outstanding Scientists) ................................ 112
Theodore Roosevelt and Hyman Rickover (Man in the Arena) ....................... 112
K'ung Chi (Virtue Its Own Reward) ............................................................... 112
Confucius (Of Thoughts and Hearts) ............................................................. 124
Ralph Waldo Emerson (Truth and Repose) ..................................................... 124
Albert Camus (Importance of Limits) ............................................................. 124
Mike (It Works) ............................................................................................. 124
Dr. Eliyahu M. Goldratt (Everybody Likes Testimonials?) ............................. 146
Charles Dickens (Best and Worst of Times) ................................................... 170
Stefani, Stefani, and Kanal (So Many Different People) ................................. 170
Confucius (People Alike, But Different) .......................................................... 170
Ralph Waldo Emerson (That Gleam of Light) ................................................. 170
Moms and Dads (Of Practice and Perfect) ..................................................... 192
Yoda (Do or Do Not) .................................................................................... 192
Nike (Just Do It) ........................................................................................... 192
Hermann Hesse (Allegory of the Stone) ........................................................ 192
Richard Bach (Blue Feather) .......................................................................... 192
Socrates (Plan for Finding Natural Goal?) ..................................................... 192
Sarah McLachlan (Unjaded by Their Years) ................................................... 192
Stefani and Kanal (Not Important to the Plot?) ............................................. 242
Stefani and Kanal (World Go Round) ............................................................ 242
Nike (Just Do It) ........................................................................................... 242
Albert Camus (Thought At the Meridian) ...................................................... 242

# List of Figures

1.1   The Theory of Constraints TOC Management System ............................. 27
1.2   A System Visualized as a Box ........................................................ 33
1.3   A System Visualized as a Box, a Machine, Producing Throughput .......... 33
1.4   The Three Fundamental System Measurements ................................... 34
1.5   Seeking Among Projects to Focus on ... Easy as 1-2-3 ............................ 38
1.6   Improvement Curves Showing that the Choice of Projects Matters ......... 39
1.7   Moe, Larry, and Curly Play Baseball (Improvement Curves) ................... 42
1.8   Five-Step Focusing Process ........................................................... 44
1.9   Current-Reality Tree (CRT) Management Process and Diagram ............. 57
1.10  Current-Reality Tree (CRT) for Typical Factory Scheduling Situation ..... 57
1.11  Ethan Hunt's List of 6 to 12 Undesirable Effects (UDEs) ........................ 59
1.12  Ethan Hunt's Initial Current-Reality Logic Tree .................................. 60
1.13  Future-Reality Tree (FRT) Management Process and Diagram ............... 62
1.14  Ethan Hunt's Draft of a Future-Reality Tree Trunk Detector .................. 66
1.15  Ethan Hunt's Second-Draft Future Tree ........................................... 67
1.16  Evaporating-Cloud (Conflict) Tree Management Process and Diagram ...... 68
1.17  Diagram of Ethan Hunt's TOC Evaporating-Cloud Tree (ECT) ............. 69
1.18  Ethan Hunt's New TOC Future-Reality Logic Tree .............................. 70
1.19  Ethan Hunt's New TOC Current-Reality Logic Tree ............................ 75
1.20  Prerequisites Tree (PRT) Management Process and Diagram ................. 76
1.21  Ethan Hunt's List of Obstacles and Intermediate Objectives .................... 77
1.22  Early Draft of One of Ethan Hunt's TOC Prerequisites Trees .................. 78
1.23  TOC Transition Logic Tree (TT) Management Process and Diagram ...... 81
1.24  TOC Transition Tree (TT) To Identify Claire Phelps as Friend or Foe ........ 83
1.25  TOC Throughput Value Added (TVA) ............................................. 99
1.26  The Ingredients of Return on Investment (ROI) in Order of Priority ........ 101
1.27  TOC Decision-Support Architecture ............................................... 105
6.1   Detective Columbo's TVA-O-OE Chart ........................................... 222
6.2   TOC Current-Reality Tree (CRT) .................................................. 233

# About APICS

APICS is the Educational Society for Resource Management, a 70,000-member professional society based in Falls Church, Virginia. The APICS vision is to inspire individuals and organizations toward lifelong learning and to enhance individual and organizational success. The mission of APICS is to be the premier provider and global leader in individual and organizational education, standards of excellence, and information in integrated resource management.

The APICS specific interest groups (SIGs) develop the society's body of knowledge. The APICS Constraints Management Specific Interest Group (CM SIG) is dedicated to collecting, disseminating, generating, validating, and disseminating educational resources in knowledge areas associated with the evolution of the Theory of Constraints (TOC) management philosophy.

APICS and CRC Press co-publish a series of TOC publications. APICS itself publishes, among many other things, *APICS — The Performance Advantage* and the *Production and Inventory Management Journal.*

APICS can be reached by telephone at (800) 444-2742, ext. 350, or (703) 237-8344, ext. 350; by fax at (703) 237-1087; or at http://www.apics.org.

# The Author

Tom McMullen is the founding chairman of the APICS Constraints Management Specific Interest Group (CM SIG) and is advisor to the APICS/CRC Press series of TOC publications. Tom is a graduate of the U.S. Naval Academy (BSEE, 1974), the Rickover nuclear Navy (U.S.S. Bainbridge), Harvard Business School (MBA, 1981), and the Goldratt Institute's Theory of Constraints (TOC) education programs. With over 20 years of experience assisting organizations with improving their strategic focus and operational performance, Tom has led major successful projects to restructure and reengineer the work, processes, and systems of organizations in a wide range of industries. Over the years, Tom has served as an employee, consultant, or educator for organizations such as the U.S. Navy, Mobil Oil, Cabot Corporation, AT&T, Advanced Micro Devices, EG&G, Dover Corporation, General Electric, Federal-Mogul Corporation, Cincom Systems, Thru-Put Technologies, and SAP.

Tom worked with Dr. Eliyahu M. Goldratt, the inventor of TOC, during the years of conceptual development and field experience that led to the book *The Haystack Syndrome: Sifting Information from the Data Ocean.* He led the first successful implementation of manufacturing software based on this book in April 1991 at a division of ITT. This implementation was the subject of a presentation at the 1995 APICS Constraints Management Symposium in Phoenix.

In 1993 and early 1994, Tom led the development of a semi-custom (not involving packaged software) drum-scheduling and buffer management system for a division of EG&G. This project was reported at the 1994 International Society for Systems Improvement (ISSI) conference at Fort Walton Beach, Florida. In 1995 and early 1996, serving in the capacity of Vice

President and Region General Manager for Thru-Put Technologies, a TOC software company, Tom led the sales, implementation, and technology planning in connection with Thru-Put's landmark TOC systems project at a large automotive supplier's plant in Ohio.

Tom is the author of *Introduction to the Theory of Constraints (TOC) Management System*, a selection in the new APICS/St. Lucie Press TOC series (APICS item #03521) and is an advisor for the series. He is also a contributor to the new *APICS Selected Readings in Constraints Management* (APICS item #05021), the 1996 and 1997 APICS Constraints Management symposia proceedings (APICS item #01558 and #01585, respectively), and the 1996 and 1997 APICS International Conference proceedings (APICS item #04018 and #04017, respectively), and was interviewed by *Entrepreneur* magazine (July 1996).

Tom is a member of the Boston chapter of APICS, the Internet Society, the National Association of Corporate Directors (NACD), and the Essex County (New Jersey) Mental Health Association. He is a member of the APICS Board of Directors, serving as the Vice President for Education-Specific Interest Groups. Tom is an independent educator and consultant based near Boston and can be reached via http://www.tbmcm.com.

# Acknowledgments

To acknowledge everyone who helped make this book possible is a very tough challenge. So many people have provided very generous support, sometimes tangible, sometimes intangible. What follows is a list of individuals and groups who helped create the environment in which I would be asked to write such a book, or helped comment on the various versions of the manuscript, or provided a key insight, and/or helped — whether they were aware of it or not — by creating great music, movies, or other inspiration over the years that helped keep me sharp, keep my energy high, and keep me on course, especially during the long process of writing, revising, and checking on things for this book. I'm sure I've left someone out, so I apologize in advance. Nevertheless, here goes:

Annifrid Benny Bjorn & Agnetha, APICS staff, APICS Boards of Directors, APICS SIG committees and members, Debbie Adee, Karen Alber, Mark Allen, Ted Almstedt, Jay April, Joe Avveduti

Richard Bach, Michael Barnes, Jerry Batt, Dave Bayley, Coach Bilderback, John Bishop, John Blackstone, Kathy Blazek, Phyllis Boyd, Juris Brandts, Cindy Braun, Robin Broadhurst, Howard Brown, Joe Buchanan

Don and Shirley Childs, Roland Christensen, Chuck Christenson, Battee Clan, Marta Clark, Richard Clark, David Bowman, Marty "Budman" Burke, David Cahn, David Carey, Charpie Clan, Condakes Clan, Michael Clark, Oded Cohen, Tom Cook, Angelo Costellano, Jay Cowles, Jim Cox, Regina Cromwell, Steve Cromwell

Steely Dan, Guy Daubert, Sam Decrispino, Bill Dettmer, Wendy Donnelly, Peter Drucker

Diane Easley, Rick Eberlein, Sylvia Ebbotson, Bob and Deborah Elder, Dick Esrey, Arthur Egendorf, Werner Erhard

Steven Farber, Fleetwood Mac, Flight Attendants Worldwide, Jeff Fort, Sarah Fortener (my favorite book designer), Kevin Fox, Loretta Fox, Robert Fox, Bill Frey, Terry Frick, Grant Fulkerson

Bill Garrison, Gas Lamp Restaurant, J. Giels Band, Drew Gierman (my favorite editor), Eli Goldratt, Natalie Graber, Sanjeev Gupta, Cheryl Gustafson

Judith Halevy, Ronni Halevy, Harriet Harris, Michele Angeles Hawley, Robert Hayes, Heart, Wilson Heflin, Jeanine Hendree, Jimi Hendrix Experience, Sue Hollo, Mel Host, Dale Houle, Tracy Burton Houle, Jim Hubbard, Lee Hubner, Ethan Hunt, Richard Huttner

Deidre Keenan Jacob, Seth Justman

Jamie Kelly, Larry Kelly, Sarah Suprenant Kelly, Wayne Kennard, Mary King, Alex Klarman, Patricia Kutt

Donna Lahner, Peter Langford, Carolyn Lea, Princess Lea, Dick Ling, Harry Lipp, Lynch Clan

Naomi Madlem, Jerry Maguire, Gene Makl, Wendy Mann, Ralph Marshall, Sylvia Marshall, Jim Mayes, Dennis McClellan, Don McGhay, Sarah McLachlan, McMullen Clan, Carol McMullen, Leo McMullen, Rick McMullen, Robert McMullen, Tom McMullen, Sr., Uta Meincke, Alex Meshar, Andi Miller, David Miller, Linda Miller, Bob Mitchell, Joni Mitchell, Karel Montor, Moody Blues, Avraham Mordoch, Alanis Morisette, Ken Moser, Madame Moser, Alex Mulholland, Peter Murphy, Ermin Murrain

Paula Neely, Sue Neff, Chuck Nelson, Mel Nelson, Paul Neri, Rob Newbold, Bruce Newell, Andy Nicoll, No Doubt, Peter Noonan, Selim Noujaim, Donn Novotny

Betsy Olmstead, O'Neill Clan

Marcia Pasquarella, Steve Pauker, Andy Pease, John Peskuric, Tom Peters, Rex Peterson, Al Petrillo, Claire Phelps, Pink Floyd, George Plossl, Police, Jim Pope, Al Posnack, John Proud, Provenzano Clan, Carol Ptak, Joanne Puerling

Pat Quigley

Arnold Rabin, Gary Rancourt, Jeff Raynes, Mark Reich, Hyman Rickover, Maury Riehl, Chip Roadman, Bill Roppenecker, Betsy Ross, Dick Rotnem

Jerry Sanders, Ralph Santoro, Lisa Scheinkopf, Greg Schlegel, Buddy Schulte, Wilma Selenfriend, Elizabeth Shea, Mike Sheahan, Paul Sheahan, Joe Shedlawski, John Sheldon, Frank Short, Nancy Siefert, Luke Skywalker, Gracie Slick, Gary Smith, Paula Smith, Smith Brothers, Tom Spangrud, Mike Spencer, Wayne Steedman, Bob Stein, Peter Stewart, Sting, David Strickland

James Taylor, John Tech, Tim the Toolman, Cheryl Tomlinson, Glen Treicher, Nan Trent

David Upton, USS Bainbridge FT Gang, USS Bainbridge Reactor and Electrical Gang

Dick Vancil, Dutch and Merilee Vandervort, Mike Ventura, Carlo Vernieri, Tom and Virginia, Bob Vollum, Bob Vornlocker

Barbara Wallace, Don Wann, Dave Ward, Frank Ward, Steve Watson, Dick Welch, Jim West, First Parish in Weston, Lynn White, Pam Wilson, Henri Wingfield, Limor Winter, Tom Wintle, Renee Zellweger, Jim Zortman

## Success!

"I love it when a plan comes together!"
≈ Hannibal Smith, *The A Team*

## So Let's Build on It

"To laugh often and much; to win the respect of intelligent people and affection of children; to earn the appreciation of honest critics and endure the betrayal of false friends; to appreciate beauty; to find the best in others; to leave the world a bit better, whether by a healthy child, a garden patch, or a redeemed social condition; to know even one life has breathed easier because you have lived. This is to have succeeded."
≈ attributed to Ralph Waldo Emerson

## Remember, All Those Little Things Do Matter

"Things can be broken down
In this world of ours
You don't have to be a famous person
Just to make your mark
A mother can be inspiration
To her little son
Change his thoughts, his mind, his life
Just with her gentle hum"
≈ Gwen Stefani, Eric Stefani, and Tony Kanal (of No Doubt),
"Different People" on *Tragic Kingdom*

# Introduction: A Physics of Anything[1]

## Book's Purposes

The purposes of this book are to:

- Introduce an important new expression of management science known as the Theory of Constraints (TOC)
- Organize several proven TOC principles, processes, and generic solutions ("TOC best practices") into a TOC management system
- Provide a compact source of "due diligence" information and "how to" instructions for those who are beginning their study and initial use of TOC
- Enable directors and executives of corporations, not-for-profits, and other institutions to initiate and guide applications of TOC and its management system
- Describe the TOC throughput value-added (TVA) financial management system, including its important new operational definition for true economic value added
- Highlight the TOC management system's suitability for generating superior degrees of short- and long-term profitability, employment stability, and stakeholder loyalty
- Weigh in on restoring the notion of "making the world a better place" as an immensely practical, satisfying, and fashionable pursuit that everyone can enjoy.[2,3,4,5,6,7,8]

# What's All the Excitement About?

## New Management Science

The Theory of Constraints is a management science invented by Dr. Eliyahu M. Goldratt, a scientist, physicist, author, educator, and consultant. Since the mid-1970s, Dr. Goldratt has used scientific methods to create concepts in management which have proven to be of great value to industry. He has encouraged and inspired others to use scientific thinking methods in their professional and personal lives. He has led and encouraged efforts to apply scientific thinking methods to areas outside the traditional "hard" sciences. These have so far included the disciplines of general management, manufacturing management, manufacturing information systems, management accounting, day-to-day managing skills, administration and content of education at various levels, and several aspects of the relationship between thinking and health. These are areas in which the thinking methods of science have not always been used, or — at least, arguably — have rarely been used effectively.

Finally, Dr. Goldratt has invented his own expression of the scientific method, the structured TOC thinking processes (TP). These thinking processes take the form of the family of TOC "logic tree" management processes and diagrams. These tools make the scientific method more understandable and practical, which makes the science approach more effective for day-to-day use, by many more people, in all walks of life, all over the world. Examples of non-industrial applications by individuals include a successful attempt to earn a place on an Olympic swim team, rapid improvements in Little League baseball batting performance, dramatic improvements in conflict management in high schools, and a variety of breakthroughs in mental and physical health. Since the scope of application in any area of is life affected by the way people think about it, the Theory of Constraints — and especially its structured thinking processes — may usefully be considered to be *a physics of anything.*[9]

## TOC's Version of the Scientific Method

Dr. Goldratt did not invent the scientific method. On the other hand, it is possible that — in his capacity as an educator and acting as a physicist applying the methods of physics to physics itself — he has, in TOC, articulated and refined the scientific method more clearly and more fully than other practicing scientists. It is possible that he has done this better than other students, historians, and theorists of the methods used in science.[10] Neither my qualifications nor my research to date have allowed me to make a final judgment on this specialized technical issue within the scientific peer review community. Not yet, anyway. All things in time.[11]

Meanwhile, there are several considerations, based on my 45 years of living in and studying many very different walks of life and my 25 years of working in organizations of all shapes and sizes, that I *am* qualified to address. So let's start with those.

### These Ivory Tower Debates Just Don't Matter in the Field

First, it doesn't matter if Dr. Goldratt turns out to be first or best in the area of articulating or refining science methods *within* his peer group of the community of scientists and physicists. This is because his primary — and possibly historic[12] — contribution has been his effective application and promotion of science methods *outside* that important, but still comparatively small, world.

How do I know this? Because, for one thing, millions of practical people around the world have read Dr. Eliyahu M. Goldratt's best-selling books — *The Goal, It's Not Luck,* and *Critical Chain* — and become fiercely motivated to make practical, constructive, and win-win changes in the world around them.[13,14]

## Good Problem To Have?

### What's Gotten into These People?!!!

#### They're Suddenly Demanding Better Solutions

Sometimes, at least at first, it seems as if nobody around them knows what's gotten into them ... not their bosses, colleagues, families, or friends. They seem no longer as patient with the traditional compromises being made within relationships and situations. Instead, they are likely to be seeking more fundamental "simple, natural, and powerful solutions" to the problems and opportunities at hand. *Why are they doing this?*

Because they see a new and more effective approach to managing. To be sure, the new approach shares some characteristics with the old. There's no intention here to "throw the baby away with the bathwater". The new approach especially shares some characteristics with the more *intuitive* styles of past managements. Still, it is fundamentally different even from those. Here's why: TOC makes those highly effective, yet implicit, intuitive processes of old *explicit!* Making previously intuitive processes explicit puts them within reach of more people, and more teams of people. This creates fundamentally new methods of day-to-day communication and managing. These produce better results, more quickly and on tougher problems, than the old ways.

## But They're Also Delivering Better Solutions

They're willing to dig a level or two deeper into the nature of current situations, and into the consensus view of plans. They formulate views, principles, and plans that engender more widespread cooperation and are of more lasting value. They do this even in the face of increasing complexity and an increasing pace of change. *How are they doing this?* Read on.

## Perception of Possibilities

The readers of *The Goal, It's Not Luck,* and *Critical Chain* have been introduced to the Theory of Constraints. They now know from first-hand experience that an understanding of TOC dramatically and permanently alters the active person's practical day-to-day calculus for deciding which desirable outcomes should be pursued and which ones abandoned. They know from experience in using TOC that some objectives previously classified (and set aside) as "very valuable and important, but just not practical" are now viewed as outcomes that can and should be undertaken and accomplished — starting today!

These people see the possibility for things to become substantially better. They care about their families, their friends, their neighbors, their various institutions, the world, and themselves. As a result, they find it difficult *not* to agitate for a consensus among the people in their lives to cause the new approach to have a very widespread impact.

## Confidence and "Do It Now!" Culture Restored

Some of these people had lost confidence and grown bitter when their simpler "do it now with quality" approaches — that once brought them consistent success — failed them in more complicated times. In TOC they see a powerful return to constructive "do it now"[15] ways. The "it" that they now do — the aggressive and rapid building of TOC logic-tree structures — feels awkward at first. However, they quickly blaze new pathways within the new tools. They recapture the flavor and momentum of their traditional "bias for action". The action proceeds, but more safely for all concerned and more effectively.

They also see a new definition of "quality" (maybe even "Quality"[16]) that serves them well when creating effective combinations of practical circumstances. In TOC they see tools that channel their energies and dedication into winning directions. They see tools that paradoxically meet their more complex world with simpler solutions. This is a particular type of simpler solution, however, that accomplishes much more complex outcomes. Because there are fundamental reasons that the prior forms of operational and cultural simplicity will never return, this is a good thing.

## Old Guard and New Guard: New and Powerful Synthesis

Some of these people had felt the need to decide between what they viewed as — under the assumption of existing management methods and tools — two mutually exclusive personal and professional choices. One group, facing the need to succeed, or maybe even to survive, chose to become the "old guard", disdainful of the new ideas and people, a drag on the natural pace and process of change. Why? Because the new ideas and people were not only changing the *forms* of things, but also the *spirit* and *intent* of things. If it couldn't be stopped, they argued to themselves, it had to be slowed. Plus, their age and experience had taught them how to slow things down with only small expenditures of time and energy.

The other group chose to become a self-described and self-congratulated "new wave", disdaining without adequate consideration the existing traditions and the older people. For a while, this group dangerously assumed that the less articulate were also necessarily less wise. They thought of themselves as the only hope, the only champions of change. Their motto: The older people and the less articulate know nothing of real value. They bring "nothing to the party". Their resistance to change is due to laziness or wrong "mindset".[17]

Both groups, upon encountering TOC, have recognized that it represents and creates the synthesis of (1) the spirit, wisdom, and good intent of the old, with (2) the energy, creative/destructive will,[18] and the good intent of the new.

## Is it any Wonder?

We are dealing here with a timely new and important expression of timeless ideas, a system that provides practical pathways through the large and small practical issues of day-to-day life. Is it any wonder then that large numbers of people are learning about and becoming powerfully motivated by this thing called TOC?

## Problem or Opportunity?

This TOC phenomenon, and its associated management system, is either a problem or an opportunity, depending on how boards of directors and senior managers of organizations work with it.

### Problem?

It's a problem when employees inspired by TOC try to tackle problems too big or too subtle to be tackled without the understanding, coordination, and support of the board and senior management. Some of the case studies mentioned in Chapters 2, 3, and 4 are this kind of TOC management system implementation. Excellent and remarkable results are often still achieved, but with unnecessary degrees of personal heroics and unnecessary side effects for the individuals and their organizations.

### Opportunity?

It's an opportunity if managements can see TOC's framework of systematic application of common sense as their structured approach for having large numbers of well-intentioned and practical people shape their organizations into sound circumstances of legitimacy, leadership, and success.

### This Book Helps Executives and Directors Sort it Out

Until now, no compact and efficient opportunity has existed for directors and senior managers to understand enough of the right things about TOC to make a thoughtful evaluation. The purpose of this volume is to provide such an introduction to TOC and its TOC management system.

# What Is Your Organization's Foundation?

Here are questions for board members, executives, and mangers to consider.

## Management System's Foundations

What are the foundations of my company's current management system? Where did those foundations come from? Do we know *why* they have worked? Do we know whether these foundations allow a wide range of alternatives or impose important tactical or strategic limitations? Would my organization be on a more solid and flexible foundation of process, measurement, and culture if we gradually (or suddenly) made the integrated TOC approaches our standard management practice?

## Infrastructure Supporting Growth

Are "growth" strategies yet another distraction from the industrial world's only appropriate priorities — namely, cost-cutting, downsizing, and reengineering? Or is "growth" here to stay? If "growth" is here to stay, are there reasons to believe that the TOC management system and its throughput value-added financial management system is the best infrastructure for fostering business growth, operational and employment stability, and superior return on investment?

## Global Standards and Timing

Is it reasonable to expect that TOC's integrated principles, processes, and systems will become worldwide standards within the management, manufacturing management, management accounting, and manufacturing information systems professions? If so, when and how do we get into motion?

## "Knowledge Worker" Productivity

Is it possible that industry has not yet arrived at the most efficient, effective, and cumulative way to approach everyday individual and organizational thinking, productivity, and learning? Are the TOC structured thinking processes (the "logic-tree" management processes), as one leading and well-placed professor of accounting stated, "the most important intellectual achievement since the invention of calculus"?[19] Or is it impossible that something so simple — even if systematically applied — could be that much of a big deal?

## Integrated System vs. Happenstance Sub-Systems

How is it possible that a single new integrated management system could be better in every way than the typical collections of traditional management sub-systems that have evolved in companies?

## The Mind of the Strategist

If my competitors were to adopt — or have already succeeded in institution-alizing — one or more of the elements of this TOC approach, what important and cumulative advantages are they building?[20]

# Blasting the Constraints on Using TOC

### Who Should Read This Book?

### The Thoughtful and Practical "Managers" in Life

This book should be read by "managers" in all walks of life. This includes those who are self-educated as well as those who have used schools to inform their views and operating methods. These may be managers of their own personal shift's worth of work on a factory drill press, managers of the sweeping of their own individual portions of an office facility, or managers of tens of thousands of employees in manufacturing, service, government, military, or not-for-profit organizations. All of these thoughtful managers understand, at least intuitively, that organizations and other institutions must, at least to some extent, be supported in the hearts and minds of their members in order to prosper smoothly. They understand that organizations must, at least to a certain extent, make sense both internally and within their social context. They will see in TOC some of the pieces that have been missing from otherwise sound and successful organizational life and will recognize the TOC management system as a comprehensive and practical means for putting those pieces to work.

## Directors and Senior Managers Included

Though intended for the much wider audience described above, this book is addressed primarily to the company director or executive who must decide to commit his or her organization to the program of education and implementation required for using the TOC management system. This is because, unless directors and executives understand the nature and value of TOC, all the other managers — from the mailroom to the boardroom — will face needless levels of difficulty and will undertake unnecessary risks of being misunderstood in their attempts to introduce the several paradigm shifts of TOC.

This book reconciles the need to keep the material sufficiently general to appeal to a large and partly high-ranking audience with the need people will have for more detail about specific TOC aspects. It does this by showing where to find the additional information people will need to understand and use their particular portions of TOC (e.g., reference to books and workshops from APICS concerning TOC factory scheduling for foremen and supervisors of manufacturing departments).

## Breaking TOC's "Due Diligence" Constraint

In other words, this book is designed to satisfy the needs of the director and executive for a basic "due diligence" investigation of TOC. It tells what TOC is, where it came from, why people like it, who has used it, and where to go to get more information. It tells individuals how to get started with TOC and how to approach company-wide applications of the TOC management system.

Until recently, when the prudent and thoughtful director, executive, or manager heard from people inspired and motivated by TOC's principles and practices, that manager was not able to check this TOC thing out using a reasonable amount of time and effort. Since TOC has been surrounded with some controversy — *good*, not *bad*, controversy (there's a difference[21]) — it was difficult to check out. I know, because I first tried to do it when I read *The Goal* in early 1989 and ran into dead ends. This difficulty has had several bad effects. One, motivated people viewed their bosses and peers as unwilling to try new ideas. Two, bosses and colleagues felt the TOC enthusiasts were being unreasonable. Worst of all, the organizations didn't gain the benefits of TOC. Fortunately, in large measure due to the fine work of professional societies such as APICS, the Institute of Management Accountants (IMA), and the ASQC Quality organization, the "due diligence" problem has been solved. That is, *if* you know where to look. This book shows managers where to look.

## The "Airplane Book"

The publisher and I have nicknamed this volume the "Airplane Book". It is designed to allow busy executives and professionals to "get their arms around TOC" during the course of the outgoing and return airplane rides and maybe one or two hotel nights of a typical business trip.[22]

# Let's Get This Show on the Road!

It is the purpose of this volume to make it substantially easier for people to understand, explain, and use, this phenomenon called the Theory of Constraints and the TOC management system. Now, let's put it to work!

## Let's Use TOC in the Workplace

The benefits to industry of the TOC management system are clear and are discussed in this volume. Let's put TOC to use in more organizations, put it to better-understood and better-positioned use in more companies, and hasten the arrival of TOC to its position as one of the global standard management best practices.

But there's more. People who gain skill in TOC in the workplace still have those skills when they go home. They will also still have them when they volunteer their services at other organizations of importance to them. This opens up huge additional opportunities for positive impact.

## "Saving the World" is Back in Style

Recently, George Bush, Jimmy Carter, Bill Clinton, Colin Powell, and Ted Turner — as well as many fine leaders, both present and past, in cultures and countries too numerous to name — have inspired and reassured by sending the message, in words and deeds, that it is a good thing to work at making the world a better place.

You wouldn't think that would have to be big news. However, you may have noticed that being a "do-gooder" — a doer of good — has become unfashionable in many circles. That is silly, of course, but it is not solely a function of the cynicism of observers, their wrong perceptions of self-interest, and the human frailties of some of the doers of good themselves. It is also because, as discussed in Chapter 5, revolutions in the world's electronics, communications, and transportation technologies have irrevocably made the world a much "smaller" place. A world has been created where diversity is not just a remote theory, but the dominant reality of our time. The resultant new complexity of our age has made doing almost anything seem less simple.

The answer is *not* to do away with the obviously correct notion of lots of people (everyone, to some extent) enjoying doing some good, but to shift the context and skills used in support of the good that is being accomplished. Enter TOC to lend a hand.

### Enough Already with the "Powerlessness" Story!

We have to be done with the notion, heard all too often today, even and especially from intelligent people of good will, that individuals no longer have a voice or representation in governments or other modern institutions.

- This view makes no sense coming from the billions of people in the world who have been raised within any of the theological back-grounds centered on the reality of the power of the various forms of "prayer". Think about that, won't you please, you billions of such persons around the world?[23]

- Furthermore, this view makes no sense coming from any of the philosophical — or even anti-philosophical! — viewpoints that ac-knowledge humankind's various abilities to create and attract progress within the reality of day-to-day life. Think about that, won't you please, you many, more varied and yet still very influential such persons around the world? Use TOC to pick your spots, grant con-sent, deny consent, and get some right things done! Why not?

- Finally, this view makes no sense when we simply consider that there are *people* on both the inside and the outside of all the organizations and institutions in question. It's too easy to form an image in our minds of some other person or group of persons. It's too easy to deal with the image, which is not the person but rather something that exists only in our mind! It's too easy simply to fail to deal with the people. "We all" and "they all" ("they all"[24] being whichever people, in whichever institutions or social groups we think are unchanging and immovable obstacles to peace or other progress) have at least the potential to hear, to understand, and act on our own enlightened self-interest. Let's help each other see around our blind spots, and enjoy the process of making the world a better place.

## Making the Difference We Can Make: It's Bigger Now

Now that we have better tools for these complex times, it's high time we got back on track. As always, we'll follow the advice of the old saying: *We'll fix what we can and accept what we can't.* However, there will be three very big differences, as follows:

- First, we'll know that a fundamentally more effective set of tools is enabling us to fix a greater number of larger things ourselves.
- Second, where people of good will have different world views and cultures, these new tools will allow digging deeper to find bases for cooperation in a much wider range of circumstances. For example, people of all countries, cultures, and religions and all sides of the current laws and power structures love the world and want it to make sense for their children and grandchildren. There's more where that came from, but — if you think about it — that alone may be a lot. After all, that's the original reason many groups started doing what they have been doing for generations. Blend in a dash of the TOC "scale of importance" concept, add warmth, stir, and we may find we've cooked up something we can work with here and there.
- Finally, the new tools will provide the process and confidence we need to formulate actions that we can take as individuals and that will contribute and make a difference on the remaining larger issues. These are the issues we can't fix alone, and those that won't fix right away. In other words, the popular phrase, "think globally and act locally", takes on new meaning when we begin to use the TOC thinking processes and logic trees to help us sort out, explain to ourselves, and — where appropriate — explain to others why certain things matter. These new tools help us to articulate clearly the already intuitively known and so-called "intangible" cause-and-effect

relationships between the little and large things we do, both visibly and anonymously, and the larger good we are determined to help accomplish. Speaking of the "little things", let's not forget that it is the "simple virtues" — things like just plain "being nice" — that lead, through sophisticated chains of direct and indirect cause-and-effect, to the more important larger objectives.[25] It's time to dispense with the nonsense that "nice" is naïve, insincere, hypocritical, or an inviting target. It's time for "nice" people to step up to the plate, place their "votes", and continuously establish and re-establish those divinely simple things we all know in our hearts to be simple common sense.[26]

Is all of this true? It's as true as we say it is, and as true as we make it. TOC may not help us solve all problems right away, but it will help us make more progress, more easily, more quickly, and on more fundamental concerns. That's reason enough to make TOC something worth knowing about, and reason enough to tell everybody you know that "making the world a better place" is back in style!

## Let's Get to Work!

So let's go to work. Let's use these TOC approaches to sort out some problems and solutions and get some right things done.

## Keep Me Posted

Let me know of your progress. Have fun!

## There's Thinking, and Then There's Thinking

"Few people think more than two or three times a year; I have made an international reputation for myself by thinking once or twice a week."

       ॐ George Bernard Shaw

## What Pilot's Art Exists for the Stormy Seas of Coincidence and Life?[1]

Athenian Stranger: "My good friend ... why am I disquieted, for I believe that the same principle applies equally to all human things?"

Cleinias: "To what are you referring?"

Athenian Stranger: "I was going to say that man never legislates, but accidents of all sorts, which legislate for us in all sorts of ways ... Any one who sees all this naturally rushes to the conclusion of which I was speaking, that no mortal legislates in anything, but that in human affairs chance is almost everything. And this may be said of the arts of the sailor, and the pilot, and the physician, and the general, and may seem to be well said; and yet there is another thing which may be said with equal truth of all of them."

Cleinias: "What is it?"

Athenian Stranger: "That God governs all things, and that chance and opportunity cooperate with him in the government of human affairs. There is, however, a third and less extreme view, that art should be there also; for I should say that in a storm there must surely be a great advantage in having the aid of the pilot's art. You would agree?"

       ॐ Plato, *Laws IV*

## The Original Logic Trees?

"Things have their roots and branches."

       ॐ Confucius, *The Great Learning* (verse 3)

# 1 | What Is the Theory of Constraints?

## A Management Science

The Theory of Constraints (TOC) is a science of management. It applies the methods of science — specifically, the methods of physics — to the general problem of managing in life.

### A Physics of Anything

The phrase, "methods of physics", suggests the disciplines of cause-and-effect thinking. It also indicates an aggressive search for the minimum number of simplest concepts that provide maximum explanatory and predictive power.

*So what?*

What do you mean, "So what?"

> *Who cares if we find a "minimum number of simplest concepts"?*
> Well … you care.

>> *Actually, I don't.*
>> Sure you do.

>>> *Sorry, but I really don't care.*
>>> Oh, come on. Think about it.

*Okay, I've thought about it. Why does it matter?*
Because of bookstores.

> *Nope. Sorry. No way that "minimum number of*
> *simplest concepts" matters because of bookstores.*
> Been in one lately?
>
> Sure. What's your point?
> Let's do this. Let's take …

## … A Walk Through a Business Bookstore, Circa 1997[2]

Have a minute, dear reader? I would like to take you for a tour of your local business bookstore. It is filled with books filled with excellent ideas. But that's the problem. Too many ideas to understand, much less absorb and use. Some are important, some are helpful only under a limited range of circumstances, and some are just plain wrong. What to do? Read on.

Step right this way. I'll get the door for you. That brings us to this month's highlighted selections near the front of the store. Included in this month's specials are, let's see, *Creating the Learning Organization VII, Reengineering the Incorporation XXIII, Reengineering the Deck Chairs on the Titanic II, Management According to the Green Berets, Dilbert and Dogbert At Work in the Twenty-First Century,* and more. You know what I mean.

But that's nothing. Wait until we go to the "Business" section. First, we have leadership and management advice from Star Trek, Attilla the Hun, the One-Minute Manager, the Two-Minute Manager, and the 26 earlier editions of the classic Learning Organization and Reengineering texts.[3] And that's not all. We even have several legitimate big names such as Peters, Skinner, Bishop, Hayes, Wheelwright, Clark, Levitt, Shapiro, Covey, Christensen,[4] Reichheld, Plossl, Senge, Forrester, and Rockart. Still, taken together, there are just too many ideas, terms, and concepts competing for the same share of mind.

Next shelf is the farseeing futurists. Let's see, today we have Alvin and Heidi Toffler, John Naisbitt, Faith Popcorn, Daniel Burrus, Scott Adams, and quite a few more. All excellent sources of ideas. But so many! How do we deal with all of this?

Next stop, the strategy gurus. We have, hold on now, Michael Porter, Cynthia Montgomery, Peter Schwarz, Pankaj Ghemawat, Gary Prahalad, George Yip, Ries and Trout, the guy in California who made Silicon Valley what it is (or at least told the world how they did it),[5] Bruce Henderson, Stephanie Marrus, Fred Davis, Robert Michel, plus a whole lineup of *new* strategy experts. Don't miss Kenichi Ohmae of *Mind of the Strategist* fame. He is back informing us that, after 23 years as a McKinsey super-partner, he's at the helm of a 35,000-member political party in Japan that's working to restructure everything in sight there. They say it is a happy man who has found his work. It would seem Mr. Ohmae should be *very* happy, because he has found a very great deal of work.

Then there are the spirited sides of things: *Jesus as CEO, The Corporate Mystic,* and more. Now for the high-spirited side of things. Let's take Guy Kawasaki, for example. No, you're right. Let's *not* take Guy Kawasaki, for example.[6]

We are just getting started. Here are the several elements of *American Culture.* We have field exercises to build your awareness in *The Fifth Discipline's* field book (you can also build your biceps carrying this one around!).

There sure are a lot of concepts here. I bet it costs a lot of money to buy all these books.[7] Bet it takes time to read them, too. Now tell the truth. Don't you think people sometimes buy them but don't have time to read them? Still, the biggest problem is they overlap, conflict, contradict, and create all sorts of new categories and terms. It becomes impossible to sort them into a coherent approach to management of work and life.

One might argue that people do not learn the practice of management in bookstores. And one might be right. But one would have to ask oneself: How exactly do sincere and hard-working young individuals, who want to support themselves and their family, prepare themselves to succeed in the business world? Think about it. What are their options? Suppose they do not have the time, money, and nice legs required to get into, say, a Harvard Business School.[8] They could:[9]

- *Talk with somebody who works in a business.* I have worked in a lot of different businesses, with widely varying workplace cultures, and with lots and lots of individual management philosophies, attitudes, strategies, and tactics. I really have to be honest and say that this isn't necessarily a much more reliable guide than the walk we just took through the bookstore.
- *Talk with somebody who once worked in a business.* Ditto.
- *Go to work in a business.* This is a good move, but ditto again.
- *Go to school.* That is, take a few years off from earning a paycheck, plus pay several tens of thousands of dollars, to go a graduate business school. This works well for enough people to constitute a small fraction of the workplace population that we should be concerned with if we're to deal with this topic thoughtfully.[10] But, to be honest, this route is also subject to the "Bookstore Effect"[11] because there is

a lot of jockeying for position and much publish-or-perish activity
going on among faculties. This creates a lot of churn in the thinking.

Most leading business schools collect and create new concepts by observing what companies who are currently leading are currently doing. As a
practical matter, they visit the companies who are willing to be visited, read
about others, and draw conclusions from there. This *correlation* approach to
generating knowledge is not wrong but presents problems. Consider, for
example, the challenge of studying IBM just before the fall, then just before
the Gerstner turnaround began, and then now. These are three very different
realities to find, with three very different consensus positions as to whether
what would be found should be imitated.

To be sure, this approach keeps business school curricula somewhere in
the ballpark on the cumulative progress made by generations of management.
On the other hand, this approach does not identify and deal with industry-
wide cost accounting, logistics management, project management, or thinking process improvement opportunities in ways that are as fundamental and
mutually reinforcing as those developed via TOC's *cause-and-effect* techniques. The better solutions have arisen by looking, at least initially, at much
less data and by asking the simple question, "What is the goal?" This is the
natural place for a management physicist to begin. It is also common sense.
In fact, the high correspondence between good physics and good common
sense is the whole point of this book. But let's not do that too fast. There are
several things in life that are best done more slowly. This is one of them.

*Okay, we've been to the bookstore. What's your point?*
Don't you see the problem, the opportunity? What if we had a set of management concepts we knew were complete and effective, even though they weren't
many and weren't complex?

> *That would be an important baseline of knowledge.*
> Right. That's why your company wants and needs a foundation of
> a legitimate management science. The winds of management fad
> and fashion can then blow through bookstores and business schools,
> without having any effect on the organization. That's until and
> unless the management scientists in the organization have dem-
> onstrated that they belong there. However, few of the fashions will
> be needed, because, remember, the foundation is built by focusing
> on the minimum number of simplest practical concepts that pro-
> vide the maximum explanatory and predictive power.

*Why didn't you say that in the first place?*
I did.

*Did not.*
Nevermind.

## While We're At It Here, What's Wrong With This Picture?

Let's have a look at the index in each of these many business books. Hmmmm ... no mention of a guy named Goldratt. No McMullen, either. Interesting. No mention of anything called TOC or Theory of Constraints. Wait. Here's one. The book about the turnaround at National Semiconductor mentions the contribution of the Theory of Constraints.[12]

*I never heard of the Theory of Constraints.*
That's because, until now, none or few of the hundreds of consultants and academics who are writing about business are writing about TOC.[13] It's a chicken or egg story. Prior to the publication of this book, almost no one wrote about TOC because they didn't understand how to deal with it and felt it would dilute their own programs and marketing efforts. Still, I find the lack of coverage surprising.

*Why do you find this surprising?*
My network of professional contacts tells me that the companies who have, at one point or another, incorporated TOC into their management systems include industrial giants such as General Motors, Hannah's Do-Nut Shop, Ford Motor Company, Texas Instruments, Hewlett-Packard, Johnson Controls, Rockwell Corporation, and all the other companies mentioned in Chapters 2, 3, and 4. I have attended presentations about applications of TOC over the years from executives, managers, supervisors, trades unionists, educators, grade-school children, church leaders, church followers, grocery store chains, insurance companies, major airlines, and armed forces of major countries. That is the kind of word that gets around.

*I agree. That's interesting.*
It gets worse. I know one brilliant woman who has written a few very good books about thinking in management. I know she has been aware of Goldratt and TOC but did not mention either one. Her second book, at least the second one I saw, pokes fun at management and improvement techniques. And you know what? She didn't even bother to poke fun at TOC! *That* is hitting below the belt. It's one thing to criticize and make fun of something.

But to ignore it! That's *really* awful. I just hope she includes TOC in her revised edition.[14]

*That's even more interesting. It's not clear what is going on here. If TOC is really a physics, maybe the physicist who invented it slaughtered a lot of professional people's pet ideas.*[15]

Good point. TOC *is* a physics. In fact, it is a ...

## ... Physics of Performance Improvement

Organizations and individuals often seek improvements, changes for the better. Accordingly, TOC applies the methods of physics to the general problem of making improvements, anywhere in life.

The meaning of "improvement" is the establishment — or the progress made toward establishing — desired conditions in life. This deliberately indicates a wide scope within which TOC may usefully be applied.

It is often useful to view the realization of desired conditions as a function of the performance of people or processes. Accordingly, TOC is frequently and usefully viewed as a physics of performance improvement.

## Physical and Thinking Constraints

The motivation that organizational or individual performance can, should, or must be improved is a declaration that some greater level of performance should be within our grasp. If it is within our grasp, but not in evidence, then something — or some combination of things — must be impeding progress. Whatever impedes progress toward an objective or a goal, shall be called a constraint.

Some constraints — those with lesser impact — impede progress along only one or a few measures of limited importance. Other constraints — those with greater impact — impede progress along many lesser measures, or along at least a few measures of great importance, or both. Every situation contains many relatively lower impact constraints but only a single or a few higher impact constraints.[16] The higher impact constraints are called *core problems* or *root causes*. Since time is everyone's prime constraint, maintaining the focus of an individual or management on identifying and acting on the higher impact constraints uses scarce time effectively. Resources are focused consistently on deep causes, not squandered repeatedly on the same symptoms. Using scarce time effectively allows individuals and organizations to do more or to do better in less time, which opens up additional opportunities.

One set of TOC tools is provided for use when limited physical resources are the primary constraint on performance. This type of situation involves a physical constraint. More importantly, TOC provides general-purpose and structured logic-tree thinking processes for use in dealing with the more subtle and complicated situations where thinking alone, in its many forms, is holding things back. These situations are said to involve policy constraints.

## Scope of Physics Widened

The Theory of Constraints, together with the associated large-scale global field research projects now in progress, may be thought of as a statement that the fundamental methods of physics are valid and useful for *all* aspects of life. After all, physics has helped to find the simple, natural, and comprehensive solutions that have been there waiting within life's "hard sciences" such as chemistry, electricity, and biology. Why not seek similar benefits in the management arts, in the vitality of organizations, and in other areas which affect the well-being of individuals?[17]

## Validity vs. Truth: A Core TOC Concept and Criterion

As a management science, TOC seeks concepts that provide increasing explanatory and predictive power. This is clearly illustrated in the evolution, over a period of 20 years, of TOC concepts in factory scheduling, logistics, and industrial management, as discussed later in this chapter.

## Another Core Concept: Verbalizing Intuition

Verbalizing intuition is a process of discovering things that an individual or a group already — at some level — knows to be true. It involves working out the ramifications of the new, newly understood, or newly affirmed knowledge. It motivates actions appropriate to the new knowledge. TOC provides a structured process that invites and verbalizes intuition, thereby capturing the best of both the analytical and intuitive modes of experience.

The word *intuition* is used to refer to those universal and commonplace phenomena that are often also referred to as *sense, common sense, gut feeling, instinct,* and the like.[18] That the analytical mode of experience has value need not be demonstrated in this advanced stage of its triumph and long-standing dominance. The intuitive mode's abilities to make assumptions explicit and to generate new thinking have long been evident intuitively to many. These modes exist in some measure in every person and certainly within all groups of people. TOC makes the synthesis explicit and effective.

## Everyone a Physicist

The Theory of Constraints is designed specifically as a practical day-to-day physics for use by anyone, in any walk of life, at virtually any age. In other words, physics is far too important to be left exclusively in the hands of the professional physicists.

## Beyond Correlation to Cause-and-Effect

The discipline of the management physicist is to use cause-and-effect modes of thinking to home in quickly on breakthroughs and other improvements that are well-tailored to the unique challenges and opportunities at hand. This allows a company to pull far ahead of those who spend a lot of time (1) mired in data analysis, (2) running in place while studying and trying to copy breakthroughs that were well-tailored to other companies' circumstances, and (3) considering these things to be "managing" productively and effectively.

## Generalists Provide "Guidance" to Specialists

It is true that TOC's tools make *specialists* more effective. However, of far greater importance, is that they make the *generalist* more able and effective. This is critical because the world has been made so much "smaller", faster, and more complex by revolutions including those in the communications and transportation sciences. It has become inescapable to recognize that every specialized profession in life is far too important to be left solely to its own specialists. This includes those whose specialties lie in creating personal fortunes. The TOC approach facilitates the synthesis of the wisdom and intuition of the generalist — people like you, people like me — with the specialized knowledge within each profession.

## Win-Win Solutions

Use of TOC dramatically increases the range of circumstances over which practical, commonsense, and "win-win" combinations of policies and actions can be formulated and successfully implemented. By placing the generalist view over the specialist and by synthesizing intuitive and analytical experience, TOC creates practical and common-sense combinations of policies and actions that are in harmonious balance and natural sequence, in any situation. If such an ideal in balance and sequence is attainable, then attaining it will be a good and valuable thing. If the ideal is not attainable, having a view of it will still draw life powerfully in its direction, and that will be good as well. Dealing directly with the thinking in important and complicated situations provides opportunities to widen the scope of interests served by breakthroughs. More stakeholders win, and to greater extents.

## Making Common Sense a Common Practice

The principles and thought processes of TOC create, communicate, and improve upon the common sense of any situation. People agree on definitions of "improved performance", define measurements, and establish scales of importance, all of which speed identification of constraints and sharpen the focus of planning and implementation activities.

## Starting Right Is Half the Battle

The Theory of Constraints is a physics developed by a brilliant and iconoclastic physicist who has spent over 20 years encountering and synthesizing both the large issues and the many practical details of people, policies, and systems of modern life. Consequently, TOC provides a small number of simple and powerful concepts that constitute the correct starting point for thinking and action to improve performance. The approach has worked for children's problems at home and school. It has also worked for making dramatic improvements to one of the world's largest automotive manufacturing and logistical systems.

## TOC Management System

The management system based on TOC has been used, in whole or in part, by industrial corporations, service companies, not-for-profit organizations, military and intelligence operations, families, friends, and individuals all over the world to improve performance, to analyze opponents more accurately, to create better work environments, to improve relationships, to increase the experience of satisfaction in day-to-day living, and to solve all manner of the large and small problems of life.

The Theory of Constraints and TOC management system can serve as an individual's or organization's primary approach to self-management. In this view, the other improvement approaches are complementary tools that are guided, focused, coordinated, and sequenced by the TOC principles and methods. The TOC approaches may also be used, in whole or in part, in support of other approaches to improved performance.[19]

## *TVA Financial Management System*

When applied to the management of organizations, the TOC management system includes the TVA financial management system, an approach to management and accounting that is unusually effective in focusing organizations on maintenance and growth of cash flows from (and for) continuing operations. The policies and procedures of the TVA system replace management accounting systems used for decision-making and performance measurement in manufacturing, service, and not-for-profit industries during the past 70 years. In contrast to activity-based and other allocation-based cost accounting systems and to "economic value added" approaches, the TVA financial system is both clear and persuasive to industrial employees ranging from factory employees to executives, easier to use, less prone to mistakes, and less costly to administer.

The TVA financial management system arises from a starting point articulated by the physicist who invented TOC. Consequently, the TVA financial system is the simple, powerful, and natural solution that Mother Nature always intended that the management accounting profession should use. In other words, it is just the sort of thing you would expect to be invented (discovered, really) by a talented physicist with no special reverence for prevailing best practice, intra-professional sensitivities, professional services marketing agendas, or unnecessary complexity. Just the sort of simplicity you would expect a physicist to discover in the midst of what, at first, seems to be a controversy over a *very* complex matter.

The argument for the traditional system is that it reuses the rules for external financial reporting. The answer to that is if external reporting rules are unnatural expressions of economic activity, are giving rise to mistakes that are harmful to both the economy and society, and are generating unnecessary costs, then the rules for external financial reporting should be changed. But all things in time. That's another book, for some other day.

## Management System of Choice

The Theory of Constraints, the TOC management system, and the TVA financial management system are fast becoming the performance tools of choice around the world because of their foundation in simple and fundamental principles, their comprehensiveness, and their broad appeal which ranges from tool rooms, to school rooms, to boardrooms.

## There Is a Best Way To Begin To Think About TOC

The Theory of Constraints should be understood first as a management science, a physics of performance improvement, and as a disciplined approach to verbalizing the intuition of anyone — and everyone! It can then be viewed as a large collection of TOC "best practice" solutions that were created through the use of the higher level tools of the management science. This perspective is critical if an individual or organization is to gain TOC's primary benefits and to gain even the lower level benefits safely and efficiently.

The alternative is to understand TOC first as one or more of its valuable "best practices", such as throughput (TVA) management accounting, critical chain project management, or drum-buffer-rope factory scheduling. These industry-leading procedures were invented using TOC principles and thinking processes.

The drawback of viewing TOC as one or more innovative solutions is that the individual or organization misses TOC's main point and most significant contribution — the opportunity to master the processes used to create breakthrough, yet common sense, solutions in the normal course of business and life.

## From Overview to Elements

This then has been an overview of TOC — its nature, its scope, and its two currently best-known formats, the TOC management system and the TVA financial management system. It has been our initial answer to the question, "What is TOC?" It has also been an opening indication of TOC's possibilities, many of which will be explored in this volume. In the next section, we will describe the major elements of TOC and its management system.

# Elements of the TOC Management System

The TOC management system is summarized in Figure 1.1. Its major elements (the ten shaded boxes in the bottom half of the figure) are both simple and few. This makes them candidates to become the minimum number of simplest concepts that the practical, day-to-day, hands-on management physicist seeks. But not so fast. Let's have a look first.

## Diving Into Our First TOC "Logic-Tree" Diagram

Figure 1.1 is a TOC logic-tree diagram, which is a product of management processes that verbalize — make explicit — some individual's, or some group's, intuition or common sense about a subject or situation. Later on in this chapter, we will discuss exactly what Figure 1.1 is, where it came from, what it represents, and how to read it. In Chapter 6, we will discuss how to get started using TOC logic trees. For the moment, in the spirit of practicing what will be preached in Chapter 6 about "just doing it", we will just dive in and use a few portions of this TOC future-reality logic tree in support of our overview of TOC. The principles shown on the diagram will be discussed below. All references to entities refer to the numbered boxes shown on Figure 1.1.

## TOC Systems Thinking Tools

The system concept is a very useful thinking tool for use in performance improvement because it helps to show that things that appear to be different and separate are actually one and interrelated. In the physical sciences, this helps to organize the thinking of specialists. In organizations, this widens the perspective of the people and departments who must cooperate to accomplish results. It helps people who design measurement and reward systems to understand that actions that make a department more efficient may render the overall company less effective. It helps differing viewpoints to identify, understand, and give priority to shared concerns and objectives.

## The TOC Management System

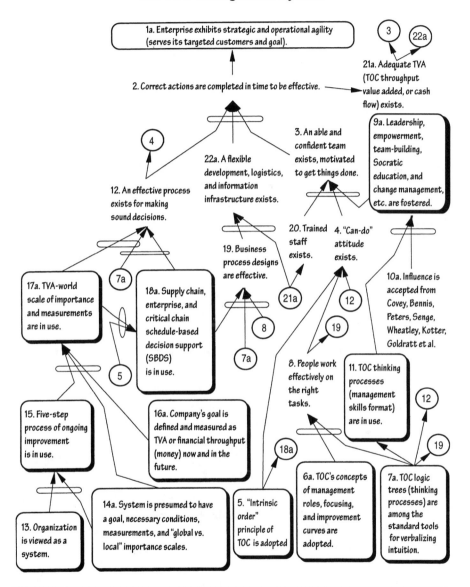

Figure 1.1. The Theory of Constraints (TOC) Management System

## Verbalizing Intuition Leads to Systems Thinking

The TOC principle of verbalizing intuition (entity 7a on Figure 1.1, bottom righthand corner) leads quickly to the need for a concept to represent a domain of activity, measurement, and decision-making. Here's why.

People frequently know from experience that certain actions are right, even though the actions may appear at first to be questionable. Their "gut feelings" tell them the actions are "good". In order to communicate the wisdom of the action, someone must verbalize these "gut feelings", which means to convert the logic underpinning those feelings from an implicit to an explicit state. This is known in TOC as *verbalizing intuition*. With this process, the riddle becomes clear. When viewed in some narrow — some local or departmental — context, the action seems wrong. When viewed in a larger — a more global or system-wide — context, it becomes clearly right. The smaller context may be an individual, group of people, or work center within a company. The larger context might be a manufacturing company, service company, or a nation. TOC calls the appropriate larger context the *system*.

## Intrinsic Order Principle Also Leads to System Concept

A fundamental TOC concept, borrowed from the classical scientific method, is the notion of an intrinsic order (entity 5, in the middle, at the bottom of Figure 1.1). The meaning of this principle was developed in the 1992 second revised edition of the popular TOC business novel, *The Goal: A Process of Ongoing Improvement*. It is productive to assume that a pre-existing order — one or more breakthrough solutions waiting to be discovered or created — exists beneath the appearance of the disorder. A useful way to think about this in most situations is similar to the "can-do" attitudes of military and other groups: If there *should* be a better way, then there *must* be a way; therefore, we should begin right away to find it and begin to act on it.[20] This begs the question, though, of a "better way" of *what*? To begin to answer this question in an orderly way, we require and we intuitively already have in our minds (have you ever noticed?) the assistance of the concepts of a system and a system's goal (entities 13 and 14a on Figure 1.1).[21]

## A Typical TOC Systems Thinking Project

Let's walk quickly through one full cycle of a TOC systems thinking project. The *system* is defined. The system's *goal* is determined. *Global measurements* of the system's progress toward the goal are established. *Necessary conditions* for continued system operation are identified and supported with appropriate measurements. *Local measurements* that support the system-level global measurements are created to guide the departments, product lines, and other supporting subgroups of the system. The TOC five-step focusing process now begins (entity 15 on Figure 1.1). This means the current state of the system is articulated, typically via use of the TOC current-reality logic-tree process and diagram. The primary cause-and-effect relationships affecting the state and the performance of the system are verbalized, made explicit. Constraints impeding the progress of the system toward its goal are identified. Other cause-and-effect processes and diagrams of the TOC logic-tree thinking processes (TP) are used to form plans that take the constraints directly into account — either by using, working around, or eliminating the constraints. The plans are implemented. The results are observed.

The system has now been altered both by the thinking processes of analysis and planning and by the actions taken in the course of implementation. In other words, the old system is gone. Some of the things that were limiting progress toward the goal, the constraints, may also now be gone. To continue to use decisions and policies created when other things were constraining is to risk having those same decisions and policies become the dysfunctional thinking, the inertia, the policy constraints that now limit the system's progress toward its goal. Therefore, it is important to rethink the new situation, to begin afresh to understand the new system, and to identify the *new* system's constraints. This iterative process is repeated for each cycle of the system's improvement process. Let's now examine each of these steps in turn.

### Defining the System

As we have seen, TOC is a physics of anything. Thus, with TOC, a "system" can be a single individual, a single organization, a reflection of a common purpose shared by several individuals, a group of organizations with some common agenda, or even some combination of these. The methods of TOC, especially the logic-tree thinking processes, can be used in all domains in life, all the "hard sciences", and all the rest.

### Goal of System

The Theory of Constraints works from the premise that every system has a goal (Figure 1.1, entity 14a). The goal of the system is

- Determined by its owners
- Measurable and should be measured
- Subject to necessary conditions

## Storyline of the Primary TOC Book, The Goal

The book, *The Goal*, is a combination of novel, love story, and business textbook that illustrates one of the industrial applications of TOC. A brisk read and a word-of-mouth bestseller, it has been handed from friend to friend and colleague to colleague, in tool rooms and boardrooms all over the world. Millions of people have gained their initial introduction to the principles of TOC via this story about how a struggling factory manager and his team saved a factory that was slated for closing.

In Chapter 4 of *The Goal*, two of the main characters meet again after many years via a chance encounter at Chicago's O'Hare Airport. Jonah is an industrial consultant, physicist, and former university professor. Alex Rogo is a former physics student of Jonah, a struggling factory manager in a large industrial corporate conglomerate, a husband with a strained marriage, and the father of two teenage children. Later, in Chapters 13 to 15, Alex will become the leader of scouts on a hike through the woods, including one named Herbie. Herbie walks a little more slowly than the other scouts, which is not unlike the bottleneck machine or department in an industrial enterprise. The hike and Herbie's assignment to beat a drum that all the other scouts can hear will become the book's central analogy for developing TOC's drum-buffer-rope (DBR) methods for synchronizing industrial operations. Alex seeks to impress his former teacher with his discussion of industrial productivity. Jonah, who knows quite a bit more about the subject than he initially lets on, listens patiently as Alex describes his problems using the language of traditional cost accounting. Jonah asks only a few pointed questions. At first, Alex thinks the questions miss the point. He soon learns that Jonah's questions *are* the point. Eventually, Jonah makes an unequivocal statement about the plant manager's fundamental problem:

> "Your problem is you don't know what the goal is. And by the way, there is only one goal, no matter what the company."

## Goal = Make Money

Later, in the famous "beer and pizza" scene of Chapter 5, Alex gets away from the action for awhile to step back and think. He has just walked out of a management meeting in which a blizzard of accounting figures are making things appear to be exactly the reverse of what they are. He asks himself what the goal of the manufacturing operation should be and considers several reasonable alternatives but arrives back to the obvious conclusion:

"I see it now. The goal of a manufacturing organization is to make money."

This should not be a surprising conclusion. After all, it takes money to pay salaries, provide benefits, advertise, buy new machines, and fund all the expenses required to keep a manufacturing company going. However, *The Goal* shows how the typical company's internal measurements, policies, and procedures are creating effects that are the exact opposite of what the company says and thinks it is trying to do — make money for all the reasons money is needed, including keeping the plant from closing. The statement of the goal is later refined somewhat to "make money in the present as well as in the future."[22]

## *Important Class of Organizations*

The organization depicted in the story of *The Goal* is a manufacturing company. This is but one example of a system (organization) whose ownership and other circumstances require cash inflows to be generated from and for continuing operations. This is a very large and important class of situations which includes all commercial manufacturing, service, and retail operations, as well as many not-for-profit, military, and government organizations.

For these systems, a great deal of coherence is achieved and many mistakes are avoided when:

1. The goal is established, as "to make more money now and in the future".
2. Money is defined in a special way that enhances operational and employment stability.[23]

To enhance stability and success, definitions for making money are organized around a fundamental money flow called TOC throughput value added (TVA). This principle is entity 16a on Figure 1.1. The TVA financial management system will be discussed later in this chapter and again in Chapter 6, by that world-famous expert in management accounting, Detective Columbo. For now, simply stated, TVA is sales less costs that vary entirely with volumes of products and services sold. This represents the money that is generated by the company from sales. If that seems obvious, it is. One needs to be an expert to be able not to see it.

As the story of *The Goal* proceeds, the main characters systematically apply the simple, but powerful, cause-and-effect thinking processes of TOC — of physics — to convert the factory from a casualty into the most profitable division in the conglomerate. Part of the solution involves side-stepping management reports and performance measures that, while designed to help the company make money, are having the opposite effect.

## Other Systems, Other Goals

That does not mean that the goal of *every* system must be to make money. At the end of the last chapter of *The Goal* (Chapter 31), Jonah takes special care to clarify that:

> "Making money is the goal for a manufacturing organization, but it isn't mine, and I don't think it's yours."

Alex tries to get some additional answers. However, continuing his maddening style of giving only questions and not answers, Jonah adds:

> "I have a suggestion for you. Think about what the goal should be."

### Different System Types, Goals, and Measurements

Different systems have different goals. Individuals can determine, or discover, their personal goals in life. Families can articulate long-term goals that align with the values they find they hold in common. Goals can be established for one-time missions such as long-range career objectives, improving a social ill, or undertaking to wipe out a disease.

### System's Constraints

Regardless of what the goal is, a *constraint* can then be defined as anything that blocks the system from accomplishing its goal. Experience suggests that constraints may usefully be organized into two categories: *physical* constraints and *policy* constraints.

Physical constraints are scarce resources such as the hours in a day, the number of production machines in a factory, skilled labor, tooling, and raw

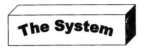

**Figure 1.2. A System Visualized as a Box**

materials. A grocery store's physical constraint may be floor space. A hospital's physical constraint may be number of beds, or number of nurses. A factory's physical constraint may be a single machine or department that operates 7 days per week, 24 hours per day, yet still cannot keep up with all the orders that have been sold. A factory's physical constraint may also be the market, or demand, for its products.

Policy constraints are all the other constraints. This category includes policies, behavior patterns, attitudes, lack of information, and everything beyond physical constraints. Policy constraints are usually far more damaging to the system than physical constraints. They are normally more difficult to identify and deal with.

## Measurements of the Goal

When the system concept is used in support of a TOC project, the system is usually visualized as a box, as shown in Figure 1.2. The system can be further visualized as a machine established for the express purpose of accomplishing the system's goal. If a particular system's goal is usefully viewed as producing something or as producing some condition, then whatever the system produces is considered to be its *throughput* (T). The usual image of this is shown as Figure 1.3. If the system in question produces money — economic value added — as one of its primary reasons for existing, then that part of the system's throughput is called financial throughput, throughput value added, or TOC throughput value added (TVA). In some contexts, it is useful to refer to this money flow as true value added.

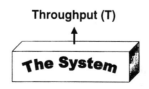

**Figure 1.3. A System Visualized as a Box, a Machine, Producing Throughput**

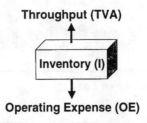

**Figure 1.4. The Three Fundamental System Measurements**

The machine will almost always require continuous feeding of some sort of "fuel" to keep it running. For the purposes of many important TOC analyses, and for the introductory purpose of this book, that "fuel" will be considered to be periodic operating expense (OE).[24] Most systems of practical interest also have an inventory or investment (I) of selected resources available within the system. These basic and generic systems measurements are shown in Figure 1.4.

## TOC's Three Fundamental Measurements

Measurements are needed in order to know if the system, or company, is accomplishing its goal of making money. In *The Goal*, Alex and Jonah work through this issue. Jonah, this time on the phone and in a hurry, says of the recommended new measurements:[25]

> "They're measurements which express the goal of making money perfectly well, but which also permit you to develop operational rules for running your plant. There are three of them. Their names are throughput, inventory, and operating expense.
>
> "Throughput [TVA] is the rate at which the system generates money through sales.
>
> "Inventory is all the money the system has invested in purchasing things it intends to sell.
>
> "Operational expense is all the money the system spends in order to turn inventory into throughput."

The Theory of Constraint's three fundamental global (system-level) measurements are discussed in more detail in the TOC book that describes the inevitable correct replacement for the problems of allocation-based cost accounting for manufacturing company decisions, *The Haystack Syndrome:*

*Sifting Information from the Data Ocean.*[26] These fundamental TVA-I-OE measurements are

- Throughput (T), also known as financial throughput and, now known as TOC throughput value added (TVA)
- Inventory/investment (I)
- Operating expense (OE)

In addition to these measurements, *The Haystack* describes a finite capacity scheduling approach that is used in support of strategic decision-making and day-to-day management of factories. Taken together, the integrated replacement for allocation-based cost accounting and factory management described in *The Haystack Syndrome* is called *schedule-based decision support.*

## If Money Is Not the Goal, Then Make Money the Operational Goal

For virtually all commercial entities, when the "goal" is defined as "money" — as established in *The Goal* and as discussed above — minimal argumentation arises. On the other hand — for virtually all not-for-profit, military, or other government entities — to define the goal or the system's throughput as money in any of its forms is to invite people to spin vigorously in circles in debates, when they could be accomplishing something useful. It is easy to see why. It is also easy to see how to exit the wasted debate. Simply define the goal and the system's throughput in whatever terms and measurements match the heart and soul of the organization's purpose, but then — if the organization also must generate cash inflows from and for a large portion of its continuing operations — define the *operational* goal as money, specifically as financial throughput, or TOC throughput value added (TVA).

Different goals require different measurements of the goal, except that in the case of the organization requiring cash flow from and for continuing operations the operational goal is always the same — TOC throughput value added (TVA).

## Scales of Importance, Part I (Global vs. Local Measurements)

When a department in a company performs so perfectly that it begins to cause the entire company to perform poorly, we have a situation where a local, or departmental, measurement is taking priority over the global, or system-wide, measurements of progress toward the system's goal. The correct condition is that the local, or departmental, measures are clearly established as less important than the global, or company-wide, measures (entity 14a on Figure 1.1). This is one application of the TOC "scale-of-importance" principle. Among the ramifications of this global vs. local principle is that it is neither necessary nor advisable to calculate allocation-based product costs for decisions.

## Scale of Importance, Part II (TVA Throughput World)

Many awful mistakes have been made in manufacturing companies by placing the reduction in inventory higher on the priority list than increasing current or future throughput value added (TVA). The same can be said for placing reductions in operating expense higher than increases in TVA. The throughput value added world (TVA world) scale-of-importance principle (entity 17a on Figure 1.1) is an explicit policy a company adopts to create a condition in which growth of current and future money inflows is given primary importance.

## Necessary Conditions

The Theory of Constraints asserts that the owners of the system have the right to establish its goal. Meanwhile, other stakeholders may succeed in establishing necessary conditions (entity 14a on Figure 1.1) that must be met for the system to continue operating. Examples of necessary conditions in the law are requirements in environmental science, workplace safety, product safety, and tax laws. There are other necessary conditions that have nothing to do with laws and everything to do with the behavior of the marketplace. Examples include customer service, product quality, and attractive pricing. Necessary conditions, with their thresholds of acceptable levels, differ from measurements of the goal, which have the characteristic of being open-ended. These are conditions necessary for empowering effective employees.

A key assumption in the TOC approach to strategy is described in Chapter 30 of *It's Not Luck,* the sequel to *The Goal.* The necessary conditions that represent the needs of a company's employees and customers carry a level of importance equivalent to the goal of making money. The importance placed on these necessary conditions requires a shift from the dysfunctional "cost world" suite of measurements and policies to the TVA world set. This, in turn, requires the well-managed TOC company to pursue growth in financial throughput value added aggressively.

### Combining Financial and Non-Financial Measurements

Much has been made in the accounting literature about non-financial measurements replacing financial measurements, and for the greater good of industry. This is not — and never was — an either/or situation. A proper construction of measurements involves the right combination of financial and non-financial measurements (entity 17a on Figure 1.1).

The "Throughput (TVA) World" measurement plan maintains a sharp focus on the primary financial measurement, the financial throughput value added, or TVA (vs. the physical "throughput" in terms of units produced) and creates a shared understanding of the cause-and-effect relationships of the non-financial to the financial measurements.

### The Ultimate Process Improvement Project: Improving the Process of Ongoing Improvement Itself

There has been a great deal of discussion about establishing a process of ongoing improvement, but there has been a lack of clarity as to what exactly is meant by the phrase and how to know when you have got a good, better, or best one. As to good ones, there are several. As to the best, let's just figure out what its characteristics must be.

### Three Questions and Roles for Management

Any management process, and certainly any improvement process, must be able to provide sound answers to the following three questions:

1. What to change?
2. What to change to?
3. How to cause the change?

The answers to these questions are often neither obvious nor easy to determine. Making these three questions an explicit part of the management agenda — and establishing the TOC logic-tree processes as the standard for answering them — is a very simple, yet very effective, executive action (entities 6a and 7a on Figure 1.1).

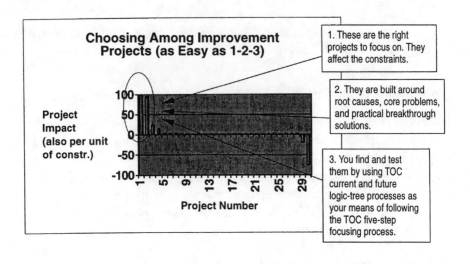

**Figure 1.5. Selecting Among Projects To Focus on ... as Easy as 1-2-3**

## Focusing the Efforts

Figure 1.5 illustrates the situation that exists in the life of every organization, individual, and group of individuals. There are impossibly many potential improvement projects. Most of the projects, those shown in the middle of the figure, make a positive contribution — both on an absolute basis and per unit of scarce resource required — but have very low impact compared with the very few high-leverage projects on the left side. That does not mean that the projects in the middle should not be done, but it does mean accomplishing them should not get in the way of identifying and accomplishing the projects with very high leverage and impact. This is similar to the concepts known as the "Pareto effect" and the "80/20 rule".[27]

The differences with TOC are twofold. First, TOC provides specific and highly effective processes for answering the crucial question: How do we identify those high-leverage and high-impact projects? Use of the five-step TOC focusing process and the TOC logic-tree processes virtually forces management teams to home in on solutions with huge bottom-line impact. The results are that TOC's processes inexorably drive toward projects that move the ratio of results to resources far beyond the "80/20" rule and into situations with ratios more along the lines of 90/10, 99/1, and 99.9/.1. There is another category of "improvement project" shown on Figure 1.5. The projects on the right side of the diagram actually cause damage to the organization.[28] The TOC management system's answer to this risk is to

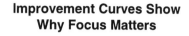

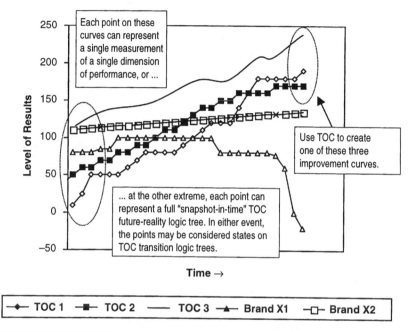

**Figure 1.6. Improvement Curves Showing that the Choice of Projects To Be Worked On Matters**

provide a companion sub-process to the future logic trees — the negative branch reservation process — to test ideas prior to implementation. The focusing diagram in Figure 1.5 should be introduced and become one of the images that people take for granted, think from, and work from (entity 6a on Figure 1.1).

## Improvement Curves

The question of how to ensure that the right problems are being addressed with effective solutions is crucial. Figure 1.6 shows why. The different rates of improvement sustained over time deliver very large differences in the absolute levels of performance, which has important ramifications for competitiveness. It is one thing to start out ahead of the competition, as indicated in the oval on the left of Figure 1.6. It's another thing to stay ahead of the competition as indicated in the oval on the right side of the figure. The difference is in the rate of improvement. The improvement rate is, in turn, determined by the organization's ability to work on root causes, core problems, and high-leverage breakthrough solutions per unit of management time and calendar time expended. On the other end of the spectrum, working on the wrong things leads to the negative outcomes and curves on Figure 1.6.

## Apply Improvement Curves to Anything

If you are involved in strategic planning for a major global industrial organization, use improvement curves to map your company's intended progress along strategic lines over time. Each point on the curve has its own associated TOC future-reality logic tree describing where the institution intends to be at that point in time. For example, the future-reality tree diagram at year three might show the new application, distribution channel, or geographic markets served then due to the steps taken during the intervening years. These trees demonstrate the wisdom of the combinations of resources and activities that are being put into place in terms of competitive and strategic positioning. The series of future-reality tree diagrams shows the development over time of the mix of human resources, capital plant, planned status of product families, expected status of base technology development projects, and projected status of the several important external trends on which the business plans of the various corporate groups and divisions are based. Designed and implemented well, these actions make it difficult for competitors to follow you quickly enough. Better yet, since some strategies conflict with others, your moves make it risky or undesirable for them to follow you at all for one of two reasons. In the first case, they walk away from their current way of serving their customers. In the second case, by trying to do it their current way *and* your new way, they do neither well and lose ground everywhere.

Maybe you aren't involved every day in matters of conglomerate business strategy, tactics, and long-term financial performance. That's okay. But you could still be on a youth baseball or softball team, either as a player or a coach. If so, you could be involved in your team's batting performance. You could define a series of points on an improvement curve that describe a state, or a level, of batting performance. You could explain the technology of batting in cause-and-effect terms that all add up to reaching one of those levels of batting performance. Let's try it.

## Sample Improvement Curve: Little League Baseball Batting

In this past season, most of the players on the baseball and softball teams were hitting well, but three players — Mo on the softball team, and both Larry and Curly on the baseball team — were having trouble getting it right. The coach decided to put them on a program of rapid improvement to help them catch up. He defined six levels of batting performance. During the course of the season, Curly started from the lowest base but improved the most. He liked the idea and picture of a vertical improvement curve. He worked with the coach to understand the cause-and-effect relationships on the future logic-tree diagram that developed and explained which combination of batting elements — in the areas of mental preparation, stance, step, focus, at-bat thoughts, and swing characteristics — inevitably had to be among the most effective class of techniques, if not *the* most effective. He paid attention and fine-tuned the "transition tree for a baseball at-bat" that converted the future logic tree's formula for excellence into a step-by-step procedure for preparing to hit the ball well each time at the plate, and getting it done.

All three players, at the beginning of the season, knew they had the choice of (1) just starting to learn and use the full internally consistent combination of superior things the coach was recommending right away, (2) picking and choosing from the recommended combination at their own pace, or (3) doing other things or doing their own things, although many of these approaches were demonstrably less effective than others. As indicated by the results in Figure 1.7, Curly committed to path (1) and Mo selected (2). Larry, who started the season as the better batter, never quite made up his mind at all, wanted to do it his own way, anyway, and defaulted into choice (3). In this season, Curly was motivated simply by the idea and the picture of the vertical performance curve. Next season, Curly and the players who saw him will be able to motivate themselves on the idea and the observed prior results.

## *Time Is the Prime Constraint*[29,30,31,32]

Knowing the answer to the question, "Which improvement process is best?", will save time in two primary ways:[33]

1.    There no longer will be a need to evaluate old and new improvement systems to make sure we are not missing something important.
2.    We can simply use the best process to get the most progress for the least time spent.

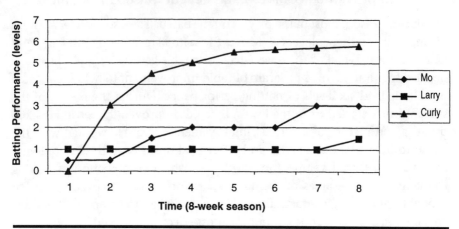

**Figure 1.7. Curly Uses a Future Tree and At-Bat Transition Tree To Establish a Nearly Vertical Improvement Curve, Which Makes for a Very Enjoyable 8 Weeks of Baseball!**

## Inherently Best Improvement Process

All legitimate improvement methods, by definition, pursue and deliver improvement. How are we to know (1) which improvement processes are valid over which range of circumstances, and (2) which process is best overall? The Theory of Constraints always asks, "What must something be if, in its category, it is inevitably to be the best that can be?"[34] What must an improvement process be if it is to be the best that a system improvement process can be?

If the performance of a system, by definition, is limited by its constraints, then why don't we find the constraints that are limiting the system the most and remove them? If we can't remove them immediately, why don't we squeeze every ounce of performance out of the system within the constraints in the meantime? When something changes that removes the original intractable constraints for us or allows us to remove them for ourselves, we start the process over again.

So that's it. That's as good as it gets. Think about it. It can get more detailed, more complicated, or more sophisticated. It can have a prettier package. You can paint it another color. But it doesn't get any better.

## TOC's Five-Step Focusing Process

In other words, the inherently best and most effective process of ongoing improvement for a system takes the form of the following five steps:

1. *Identify* the system's constraint(s).
2. Decide how to *exploit* the constraint(s).
3. *Subordinate* everything else to the above decisions.
4. *Elevate* the constraint(s).
5. *Return* to step 1. Don't let inertia become the new constraint.

Now we are ready to judge all the many and competing improvement processes against these criteria. But why bother to do that? Let's just use the criteria themselves — the five steps — and save a lot of time.

## Using the Five Steps

The five-step process (entity 15 on Figure 1.1) can be used in any situation. It is used in strategic planning, project management, process improvement, manufacturing continuous improvement, and day-to-day factory scheduling.

## General Purpose Process

For all but the most straightforward improvement situations, the five steps are used in conjunction with the TOC logic-tree thinking processes (TP), as indicated in the middle column of Figure 1.8. The structured TOC logic-tree thinking processes, which will be discussed in the next section, enable the complex problems to be handled systematically. They provide effective procedural support for this five-step process. Proper use of the TP tools make it almost unavoidable to find core problems and effective breakthrough solutions. As a result, people work on issues that deliver leverage and bottom-line impact. In the case of organizations, this enhances employee and customer satisfaction and confidence and helps ensure a healthy rate of cash flow to fund everything else that needs to be done.

## Three (of Five) Steps To Begin Lean and Agile Manufacturing

Drum-buffer-rope, buffer management, and dynamic buffering factory scheduling and management procedures deliver excellent customer delivery service, high current and future profitability, and lowest reasonable amounts of protective inventory and capacity — the recipe for superior return-on-investment, ROI — and are based primarily on the first three steps of the five-step focusing process.

| TOC's Five-Step Focusing Process | | |
|---|---|---|
| **Focusing Steps** | **All Situations (General purpose)** | **Physical Constraints (Simplified)** |
| **Step 1.** Identify the constraint(s). | 1. Use TOC logic trees to sort facts (entities) into constraints, root causes, and core problems. | 1. Identify the resource or resources that are the primary obstacles to progress toward the goal. |
| **Step 2.** Decide how to exploit the constraint(s). | 2. Use TOC logic-tree processes to decide how best to progress toward the goal within current constraints. | 2. Decide on a plan for the primary constraint that best supports the system's goal. |
| **Step 3.** Subordinate everything else to the above decision. | 3. Bring other factors in line with above decisions. Do not allow other improvement initiatives to interfere with the high priority of the above decisions. | 3. Alter or manage the system's policies, processes, and other resources to support the above decisions. |
| **Step 4.** Elevate (lift/remove/break) the constraint(s). | 4. Use TOC logic-tree processes to select constraints for removal in the current iteration and to remove them. | 4. Add capacity or otherwise change the status of the original resource as dominating primary constraint. |
| **Step 5.** Return to Step 1, but do not let inertia become the constraint. | 5. Go back to Step 1, but do not allow previous decisions made in Steps 1 to 4 to become unnecessary and damaging constraints now. | 5. Go back to Step 1, but do not allow previous decisions made in Steps 1 to 4 to become constraints. |
| **Comments** | This general-purpose process handles any combination of policy, process, and physical resource constraints. | This simplified process is the basis for TOC project management, logistics, and factory scheduling procedures. |

**Figure 1.8. TOC's Five-Step Focusing Process, the Inherently Best Process of Ongoing Improvement**

## Uncovering Hidden Capacity and Process Improvement in Manufacturing

Many improvement situations are either dominated by physical constraints, or — for some reason — are best approached via consideration of physical constraints. The righthand column of Figure 1.8 summarizes the use of the five-step process for these situations.

As indicated in Chapters 2 to 4, many companies — including several large semiconductor manufacturers — have used this five-step process, sometimes in conjunction with drum-buffer-rope scheduling, to uncover large amounts of hidden production capacity in their factories. Each time the primary constraint is identified and exploited, throughput is increased and non-value-added activity is removed. The non-value-added activity is removed first from the constraint directly, but also from all other resources and processes that affect the constraint indirectly. This allows processes to be improved with surgical precision and with high bottom-line impact for the time spent in the improvement process.

*Current-reality logic trees* are used to characterize the combination of factors (often many different policies and procedures) that are adding up to non-value-added time on the constraint. *Evaporating-cloud logic trees* and *future-reality logic trees* are used to formulate and evaluate combinations of changes to the several policies and procedures that constitute a complete and sound change. *Prerequisites logic trees* are used to break large parts of the solution into smaller and more manageable pieces. *Transition trees* are used in three ways: First, to document selected existing processes. Second, to plan and accomplish one-time changes. Third, to design and document the new recurring processes.

## New Standards for Using Time Well

To be "using time well" means that the answer to "what to change" must have come from or been verified by the use of TOC logic trees. Current-reality and future-reality logic-tree processes have clarified and verified the nature of the root causes, core problems, and practical breakthrough solutions — the three primary components of the "intrinsic order" waiting to be found within the situation. The next section will explore the TOC concept of intrinsic order.

## Intrinsic Order Principle

The ability to act effectively is of paramount importance. Many fine plans have remained on the shelf as the world passed by those who failed to take effective action.

Did you ever think about what gives rise to a confident "can-do" environment? One could take the view that a strong leader inspires this. But what is the strong leader thinking — at least implicitly — that gives him or her the ability to radiate such confidence? One could answer that it is a prior track record of doing the right thing. Fine, but was that due to being lucky and being in the right place at the right time, or because somebody who was there assumed they would find ways to succeed (maybe because necessity was the mother of invention) and *did* find those ways? Either way, there is the belief or experience on somebody's part that a way is there and can be found. So why not extract the fundamental lesson from this day-to-day observation and presume the existence of a pre-existing order, an intrinsic order, waiting there to be found?

We are often forced to assume that there is a solution waiting to be discovered. And, once we do so, we find that this presumption is valid over a *very* wide range of circumstances.

The TOC technique is to wade right into an important and complicated problem and systematically replace unnatural, illogical, or overly complex situations with the classic TOC outcome using simpler and more natural common-sense solutions.

Under the presumption that the performance of the system in question is limited by one or a few constraints, a simple and powerful solution — an attack on those constraints — always exists within the perception of complexity. The project team may not yet have discovered the path, but it is there. This "can-do" attitude creates the basis for effectiveness.

Because competition, success, cash flow, and continuing operations are matters of relative performance, think about the ramifications of this simple and arguably obvious (once one has thought about it) principle becoming a standard notion throughout various types of competitive activities. Interesting.

Does making the presumption of an intrinsic order guarantee that a breakthrough solution will be found in time to meet the organization's needs? No. Does it dramatically increase the odds of creating and accomplishing a steady series of breakthrough solutions that cause the organization to move beyond survival to a habit of thriving? Absolutely!

## Logic-Tree Thinking Processes
## Reveal Intrinsic Order

This raises the question of how best to find the intrinsic order in any situation. This is the purpose of the TOC logic-tree thinking processes. The logic-tree processes provide a systematic approach to verbalizing the knowledge and intuition of internal and external experts, in order to construct workable solutions to large and small problems.

# TOC Thinking Processes (Logic Trees)

We have established that the Theory of Constraints (TOC) is a new expression of management science invented by Dr. Eliyahu M. Goldratt, the scientist, physicist, author, educator, and consultant. We also know that, since the mid-1970s, Dr. Goldratt has used scientific methods to create concepts in management that have proven to be of great value to industry. He has encouraged and inspired others to use scientific thinking methods in their professional and personal lives. He has led and encouraged efforts to apply scientific thinking methods to areas outside the traditional "hard" sciences.[35]

It is reasonable to ask how he has done these things. The answer is that Dr. Goldratt has invented his own expression of the scientific method, the structured TOC logic-tree thinking processes (TP).

This has made the scientific method ...

... more understandable and practical ...

... for day-to-day use ...

... by many more people ...

... in all walks of life ...

... all over the world.

Take Ethan Hunt, for example. Before we move on, though, I have to give you one word of advice. If you haven't yet seen the movie, *Mission Impossible,* stop reading and get the video right away.[36] Fasten your seatbelt, watch the movie, and now use this section to sort out some of the things that went by too fast to follow. Or read this section to prepare to keep track of what's happening.

Ready? Okay, here we go.

## Mission Impossible: *The Movie*

Ethan Hunt (Tom Cruise) is in very deep trouble and now he knows it for sure. His spy operation at the embassy party in Europe has become a disaster. Of all the members of his team, only Ethan is still alive. The rest were mangled in an elevator shaft, knifed through a fence, shot on a bridge, or blown up in a car. *What went wrong?!*

Shortly afterward, in the same city, he turns for help to Kittridge, one of his bosses in the spy organization; however, that soon becomes another fiasco. Kittridge gives him the facts. The bad guys have made off with a list of spy names during the embassy disaster. *Well, that much I knew.* The spy list was a fake. *You mean this whole thing was a sham to find a "mole", a bad spy?* Someone has wired money into Ethan's relatives' bank account back in America. *What?! Wait a minute.* Ethan is the only survivor of the botched embassy operation. *What's going on here?* Add it up. Kittridge thinks Ethan tried to sell intelligence secrets. *He thinks I set my team up for the kill!* Ethan has only one choice, one chance, for escape from Kittridge's goons, so he comes flying through the shattered glass window of the restaurant and runs for cover.

But now what? There is not only a question of *what* to do, but also the question of *how* to begin to approach a problem like this.

## There's Thinking and Then There's Thinking

He could just sit or stand there and let the furious stream of thoughts and emotions flow through his experience. Ethan knows this is what many people call "thinking". He also knows that sort of activity won't get him anywhere that is safe or good. In fact, in this kind of situation, it will only kill time and cause him to be without an intelligent plan — still without a plan — when he is eventually forced to make *some* move, *any* move.

He knows he has to get beyond the chatter and the fireworks of the free-flowing thoughts and emotions, and really think. Experience has taught him that one way to get a foothold in the real thinking of a situation is to talk with someone about it. Verbalizing his thoughts, his sense, his intuition about a complicated situation — out loud to someone who also cares about him or the situation — will normally bring him back to a state of knowing he is really digging into the matter. When this happens, ideas still flow, but more usefully, more powerfully. Unfortunately, there are two problems here. One, not everyone has the skills necessary to be effective on the listening side of the type of conversation that leads people to better ideas.[37] Two, in this situation, it is not clear who would be a safe person to entrust with his thoughts.

Maybe he could just talk out loud to himself. This sometimes has the same effect of improving the state of mind and producing breakthrough insights and ideas. Sometimes this method's effectiveness seems to depend on who the person has in mind as the implicit, or sometimes explicit, "imagined" "listener" when he does it. Naturally, one concern Ethan would have about talking out loud is that it may draw attention to him.[38] He doesn't need that right now. It can also look, well, a little nuts.[39] Plus, think about it. In his line of work, Ethan has to assume every flower pot comes standard with a built-in microphone.

He knows letter writing will work. It always does. Addressing himself to other people as he verbalizes his intuition in letters routinely produces the insights and plans he needs, even though the letters themselves are often never sent. Prior to learning about TOC, verbalizing intuition via writing or speaking out loud was all he used to do. Still, he has a better idea about how to handle this particular situation.

By now, Ethan also has a lot of experience with TOC logic trees. He knows that, even if he starts with free-flowing verbal or written verbalizations of his thoughts and intuition, the heart of the situation analysis and the crux of any sound plans will express themselves in smaller numbers of clustered and precisely worded individual cause-and-effect statements. In other words, he gets where he is trying to go most quickly and most effectively if he starts with — rather than backing himself into — the format of the TOC logic trees.

## Mission Impossible: *The TOC Case Study*

Ethan knows that, when things really need to be right, it is best to gain added focus, rigor, and speed by using the TOC logic-tree thinking processes right away. That means the question is not *whether* to use logic trees, but *which* logic-tree process tool to select.

## *Tool Selection Time*[40]

Ethan's situation is the type for which it is not clear where or how to begin. With TOC, this means that one should:

1. Create a vision statement by allowing intuition to flash a series of potential new ideas, or injections into the mind for rigorous processing via TOC future-reality logic-tree processes.
2. Make a list of 6 to 12 undesirable effects (UDE) to begin forming a fresh TOC current-reality logic-tree that describes the situation.
3. Create a TOC evaporating-cloud logic tree that captures and builds from one of the obvious and powerful conflicting imperatives in the situation.

Having been shot at, nearly knifed, almost blown up, and nearly cut to shreds by flying glass, Ethan is feeling a little short at the moment on vision and brainstorms. That rules out plan (1), the future tree, at least for the moment. The cloud is often an elegant entry into the thinking; however,

Ethan's mind and emotions are positively reeling from a dozen different and conflicting imperatives. Sometimes that makes for a speedy run straight to the heart of the situation. Today, though, what feels right is to begin by facing the facts in the current situation. That means starting with — drum roll, Professor, if you please — plan (2), the TOC current-reality logic tree.

Let's leave Ethan Hunt to catch his breath for a few moments. We will use this opportunity to introduce the five major elements of the TOC logic-tree thinking processes. With that new ground under our feet, we will return to Ethan and see how he uses the thinking processes to get himself out of this mess.

## Summary of TOC Logic-Tree Thinking Processes

The TOC management system employs several management processes. Five of these processes are grouped under the category of TOC thinking processes (TP). The tool set for the thinking processes (Figure 1.1, entity 7a) has five major components, each its own management process with specific rules and deliverables (diagrams):

- Current-reality logic tree (CRT) process and diagram
- Evaporating-cloud logic tree (ECT, or EC) process and diagram[41]
- Future-reality logic tree (FRT) process and diagram
- Prerequisites logic tree (PRT) process and diagram
- Transition logic tree (TT) process and diagram

Each of these five processes has associated deliverables that include diagrams of cause-and-effect relationships which exist within situations.

These diagrams have come to be called logic trees. Here's why: Many readers of this book will be familiar with the fact that Total Quality Management (TQM) practitioners long ago discovered that their diagrams resembled the bones of fish and began to call them "fishbone diagrams". Similarly, TOC practitioners have discovered that their cause-and-effect diagrams often resemble trees, with trunks and branches, and have come to call them "trees". One of the diagrams is called the evaporating-cloud logic tree; this term arose from the fact that, as assumptions in situations were challenged and changed, the conflicts addressed in the diagrams disappeared as if they were evaporating clouds.

These TOC thinking processes — the logic trees — are designed to home in on the breakthrough solutions that already exist within a situation, solutions which are there to be found. In other words, they are intended to assist in finding the intrinsic order within a situation.

## Benefits of the Logic-Tree Thinking Processes

■ A company or individual uses TOC thinking processes and diagrams for strategic planning, policy formulation, process management, project management, day-to-day problem solving, and day-to-day management.

■ The thinking processes enable a team of people, or an individual, to create and evolve a clear common sense of a situation, proposed solutions, and plans.

■ The rigor of the cause-and-effect processes and diagrams is supported by the categories of legitimate reservations, a set of rules to guide analysis and discussion. These guidelines focus the efforts of individuals, and enable projects involving teams to gain the best thinking and contribution of *all* participants, regardless of rank or verbal communication skills.

■ These processes provide a means for systematically exploring the relevant explicit and implicit assumptions involved in an improvement project. This leads rapidly to finding and creating breakthrough solutions in strategy and other areas.

■ Clarity in individual applications and consensus in team situations are built quickly and then maintained by using prerequisites trees and transition trees to develop and measure the implementation plans.

■ Through use of the thinking process tools, companies become effective "learning organizations" by systematically evaluating popular "best practices" against their own and outside expert common sense and by inventing their own solutions and "best practices". Individuals deepen their understanding of the world around them.

■ The teams in the learning organization use the cause-and-effect diagrams (the logic-tree and cloud diagrams) to capture and document the learning organization's continually evolving know-how. This library of learning company know-how can be on paper or, increasingly, in graphical and networked inter- or intranets. Individuals develop a knowledge base.

## Let's Pay a Visit to Ethan Hunt

It looks like he has nearly finished making the list of undesirable effects he sees in his current situation. This means he'll soon be using that list to create his first draft of the associated TOC current-reality logic tree. It doesn't take very long to do this. In fact, it takes longer to write about it than to do it. Let's not disturb him yet. Let's go back to some things we know about Ethan's prior experience with TOC logic trees.

## Four Stages of Struggle, Practice, and Skill with Logic Trees

Ethan has observed that, while initial successes can be easy with these tools, struggle and practice are necessary to master their use and to gain the largest benefits. After his initial serious exposure to the logic trees and after some practice, he arrived at the view that there are four stages in the development of the skills.

### Stage One

First, and for awhile, he gained a large amount of productivity by simply having the *images* of the logic trees in his mind, knowing that he should and might try to draw the trees. This went on while he was thinking via writing free-form text and pictures and speaking out loud with well-chosen listening partners. Having the logic-tree structures in mind increased the focus and effectiveness of his thinking — even if he never wrote the tree diagrams. He felt guilty, of course, for *not* writing and refining the actual trees, but that was a force that was leading him into an arguably correct direction for future improved effectiveness.

### Stage Two

Ethan entered a second stage when he actually started to write the trees. He usually did this after the whirlwind of energy surrounding his thinking via free-flowing oral or written activities had subsided. Ethan is a results-oriented, do-it-now sort of active person. That's why, initially, his emotions made it difficult for him to nail himself down to a chair long enough for the incremental thinking required to convert the brainstorms from the verbalization activity into the rigor and precision of the logic diagrams. Though difficult, it was literally rewarding beyond his expectations. "Literally" because, having lived in one particular mode of experience of problem-solving, he hadn't really ever even wondered whether other people were experiencing and dealing with the same day-to-day situations differently or whether that made any difference to him. In other words ...

... he really didn't know ...

... to expect the feelings ...

... of becoming more deeply anchored ...

... and grounded in being ...

... via nailing down ...

... one ...

... first-order principle, ...

... fact,

... and cause-and-effect relationship ...

... after another ...

... in the regular course ...

... of handling ...

... the day-to-day details, ...

... of work ...

... and of life.

His emotions, temperament, and bias toward action — all positive things — always pulled him into action well before much, if any, real thinking got done. On the other hand, in earlier years in other environments, the sort of thinking made almost inevitable by the TOC's simple framework, procedures, and disciplines was often neither needed nor wanted. But times and circumstances have changed. This sort of thinking is now in demand.

## Stage Three

This is the stage that Ethan is in right now. When he recognizes a situation as having high importance or complexity, he anchors himself immediately by focusing in on individual logic-tree entities. The intent is to get to the heart of matters with the least elapsed time and wasted motion. It was difficult at first to shift immediately to the trees' operating mode, because the prior free-flowing oral and written styles of verbalizing the thinking not only worked but also felt good. He had taken for granted the self-expression, increased confidence, and effects on leadership and power he had derived from his prior style of verbal initiative. He gradually understood and adjusted the sources and equations of personal satisfaction. The persistence paid off. The boosts to performance — and to feelings of deepening personality, greater certainty of doing the right things at the right times for the right problems, and general clear-headed well-being — were huge.

By now, in Stage Three, Ethan knows he is very good with TOC logic trees. For simple problems, he does not draw trees at all. He "wings it" along with everyone else. For some complicated or important problems, he reverts to the free-wheeling energy of the written or oral verbalizations.

With increasing frequency, however, he turns to the logic trees first. This is partly because he gets where he wants to go more quickly and partly because it prevents him from working on the same symptoms over and over again. Every time he now returns to an area of work or life, the work with the logic trees creates two kinds of knowledge for himself: (1) knowledge and conclusions that are useful to the specific current pressing situation alone, and (2) knowledge useful to either *similar* situations or *all* situations. This also means his understanding of the meaning and value of the TOC intrinsic order principle increases in the course of his day-to-day work. But more about that later.

## Stage Four

Ethan knows he can further improve his efficiency and effectiveness — if he wants to — through further refinements in understanding the formal logic-tree disciplines and by paying more consistent attention to technique. All things in time. You sing better after singing lessons, but you do not need to know everything about singing technique to sing a lot of songs and sing them well. Same for playing tennis, golf, basketball, baseball, and virtually all sports and games. Same for TOC thinking processes.

In fact, this is what Ethan has always liked most about TOC. Someone can — and should — just dive in with any of the methods of using the TOC principle of verbalizing intuition, of systematic application of common sense. That allows a person to get satisfaction right away from the effects on deepening of personality and from the improved external results without waiting to be sure every little detail of the most orthodox advice on technique is understood.

The positive effects on the problem-solver's feelings and state of mind come from (1) moving more quickly and more effectively into successively more meaningful levels of what the solver knows are the more important problems, (2) working with the solver's full analytical and intuitive resources vs. just one category or the other, (3) getting results, and (4) steadily gaining skill by taking incremental steps toward the use and then more formalized use of the TOC logic-tree processes, at his or her own pace.

The improved results come from *explicitly* using the same improvement process that brilliant intuitive physicists, executives, engineers, marketers, salespeople, politicians, clergyfolk, and other creative practical geniuses have always used *intuitively* to deal with important and complex problems.

## Ethan Hunt Uses TOC Logic Trees?

This is the part that makes this book a lot more fun and exciting than the movie.[42] You have to understand that Ethan Hunt does not allow his use of the TOC logic-tree thinking processes to show on camera. Oh, but the results speak volumes.

A lot of TOC experts are like that. They seem to just keep pulling rabbits out of hats as if it were nothing at all, and you think you never see the logic trees written down anywhere. But the trees are there, at least in the mind's eye of the person in motion, but also scrawled down somewhere on shirtsleeves, napkins, and other fine stationery. Once someone has really studied and learned about TOC logic trees — even and especially if they have always used a similar process intuitively to get great results — they at least dimly "see", in their mind, the logic trees that reflect the progress of their thinking.[43] If logic trees are actually written for a specific situation, those trees will still be in the mind, although dimly perceived, even when the situation does not require or allow reference to the original written trees.

Now let's see what the movie would look like if Ethan had allowed his use of TOC logic-tree thinking processes to show on camera. At tool selection time,[44] Ethan elected to begin with a TOC current-reality tree process. Let's take a moment to make sure we understand what he has in mind.

## Current-Reality Tree Process and Diagram

### The Heart of the Matter

Among the wonderful phrases in the English language is "getting to the heart of the matter". We know, intuitively, what that means. But think about it again. What does it mean that a matter can have a heart? One can take this in either of two interesting and related ways. One interpretation is that the discussion has reached such a state that it touches the hearts of the human beings involved in the situation, as opposed to their remaining detached in an analytical domain. The other, as indicated by the intrinsic order principle, is that there are patterns beneath the seemingly random data, and that discussions can reach states where such patterns become clear, at least intuitively, to the people involved. When this happens during a meeting, we know it. The tone of the discussion and the tone of the relationships in the room all suddenly change a little, sometimes a lot. Some people have attuned themselves to notice this effect more than others, but there is no question about whether it happens. What happens is that the discussion reaches a heart of the matter — a root cause, core problem, or breakthrough solution — and everybody in the room intuitively knows it. The purpose of the TOC current-reality logic-tree process and diagram is to make the connections among a current situation's many symptoms, facts, root causes, and core problems explicitly clear to everyone in the room. That's *everyone*, not just the one or a few people who seem to have the knack for always managing to come up with that great insight, articulating it, and sounding sufficiently on target that we cannot really disagree, yet still not making it clear enough so that everyone really understands.

People do not really have to understand the issues in order to nod heads in the affirmative at meeting room tables, allowing them to look smarter than the other people and ensuring that the power players running the meeting think they are playing ball. On the other hand, the people *do* really have to understand the issues in order to really feel comfortable about them, to provide strong support for the views and plans on the organization's informal grapevine, to offer intelligent improvements to the ideas to increase the likelihood of success, and to avoid unnecessary mistakes during implementation of the plans.

Figures 1.9 and 1.10 are sample (and simple) diagrams of TOC current-reality logic trees. Figure 1.9 is a generic diagram which indicates that core problems and root causes, typically shown at the bottom of current-reality tree diagrams, are the causes of the undesirable effects shown at the top. Figure 1.10 is a portion of a current tree showing the typical situation in a factory prior to introduction of TOC's drum-buffer-rope scheduling and buffer management shop floor control systems. Entities 36 and 37 are caused by the lower entities 30 to 35. On the complete tree, entities describing the resultant negative effects on delivery service to customers, inventory levels, quality levels, employee morale, and current and future profitability would be shown.

## Current-Reality Logic Tree

- Current Situation

- Undesirable Effects

- Facts, Entities

- If... Then...

- Cause... Effect...

- And (linked)... Or...

- Root Causes

- Core Problems

**Figure 1.9. Current-Reality Tree (CRT) Management Process and Diagram**

## A Typical Current-Situation Diagram
### Scheduling

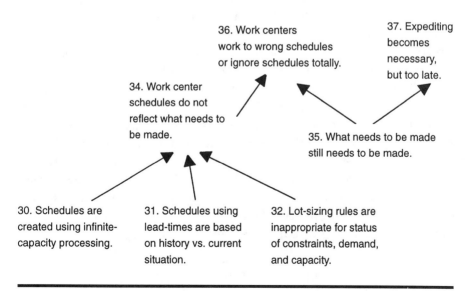

36. Work centers work to wrong schedules or ignore schedules totally.

37. Expediting becomes necessary, but too late.

34. Work center schedules do not reflect what needs to be made.

35. What needs to be made still needs to be made.

30. Schedules are created using infinite-capacity processing.

31. Schedules using lead-times are based on history vs. current situation.

32. Lot-sizing rules are inappropriate for status of constraints, demand, and capacity.

**Figure 1.10. Current-Reality Tree (CRT) for Typical Factory Scheduling Situation**

## The Facts Line Up at Harvard Business School

Here is another take on the nature of the TOC current-reality tree process and diagram. During my second year of study at Harvard Business School, I elected to take a course on Corporate Finance, or maybe it was another course on Business Policy. Though I was very intrigued by the content of the course, I just wasn't motivated to participate during the classroom discussions. I didn't have anything that just needed to be said in that course, which was unusual for me, but I did not go out of my way to make things up to interject either. Then came the term paper.

It was a complicated case about relationships among managers. Knowing I was behind on class participation, I resolved to ensure I did well on the written case study project. I sweated over each bit of data, made sure my theory of the case matched every fact, adjusting the theory many times to do this, and hit the analysis exactly right.

Problem was, during the course of the evening, I had to chase a burglar out of the apartment building and then deal with the police reports and such. By the time the all-nighter and revisions (with a manual typewriter!) were finished, one class (not mine) had already discussed the case in class. So the fellow teaching the class (there really was a reason I had not been participating) decided I had gotten a spy from the morning class to somehow leap out of the discussion to tell me the "real" solution to the case, which few people get (via, oh, I don't know, radio or telepathy?) in time for me to deal with it, type it, and turn it in (which was physically impossible, by the way[45]). But I had cracked the case so cleanly and completely that he could not believe that this guy who hadn't said anything in class the whole semester could possibly have done it himself.

Well, I did. How? I didn't know about TOC at the time. This was 1980 and 1981, and Dr. Goldratt and his colleagues at Creative Output were working on solutions to factory scheduling. The current-reality tree process and diagram as they are known today would not appear until a decade later. Nevertheless, the process I used was the very heart of the current-reality tree process. In other words, the essence of the current-reality tree process is to get a theory of the situation that orders all the key data in the situation. If I had known today's TOC current-tree procedures then, I would have nailed the case just as cleanly, but I would have taken less time to do it.

## Several Ways To Begin Current Trees

There are several ways to form a TOC current tree. All are designed to get straight to the heart of the matter by beginning, not with all the possible data,

---

**Undesirable Effects for Ethan Hunt**

UDE 1:    Everyone on my team is dead except me.

UDE 2:    Money has been transferred to my relatives' bank account.

UDE 3:    Kittridge thinks I'm the bad-guy spy.

UDE 4:    Kittridge is trying to put me in jail.

UDE 5:    My career is over.

UDE 6:    I am alone.

UDE 7:    Once the blame is fixed on me, maybe I get killed next in an "accident".

UDE 8:    Whoever really killed everybody and set me up is still out there.

---

**Figure 1.11. Ethan Hunt's List of 6 to 12 Undesirable Effects (UDEs)**

but with the minimum practical data. In other words, do not read all the books in the library to get the ten sentences you need. Use techniques that help you get to what you need more quickly and more directly. For the particular and oft-recurring situation of not knowing how or where to begin, TOC's default recommendation is clear: prepare a list of 6 to 12 things you do not like about the current situation, 6 to 12 undesirable effects (UDEs). This brings us back to our hero, Ethan Hunt.

## Building a TOC Current-Reality Tree

Ethan has begun by making his list of undesirable effects (UDEs), which are shown in Figure 1.11. Ethan now begins to apply TOC's categories of legitimate reservations (CLR) to these UDEs. Beginning with the entity existence reservation, he finds that the entity called UDE 6 no longer exists, because Claire Phelps has suddenly shown up. He removes UDE 6 from the list. However, here he begins to depart from the formal procedures. He is in Stage Three of TOC logic-tree skill and knows it. He draws logic-tree diagrams and gets a great deal of value from it, but he does not really follow — or really even know for sure — all the well thought through, finely tuned, and well-tested standard procedures for forming current trees. He wings it as far as determining which UDEs have direct cause-and-effect relationships with other UDEs and connecting them with arrows to form a cluster of UDEs. The result is Figure 1.12, Ethan Hunt's current tree. It is not a pretty picture, for more reasons than one.

Ethan's experience with TOC has taught him that the intrinsic order principle and the one-or-a-few-constraints principle are either always valid or

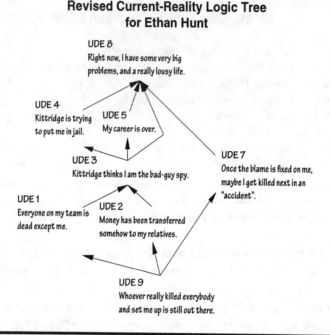

**Revised Current-Reality Logic Tree
for Ethan Hunt**

**UDE 8**
Right now, I have some very big
problems, and a really lousy life.

**UDE 4**
Kittridge is trying
to put me in jail.

**UDE 5**
My career is over.

**UDE 3**
Kittridge thinks I am the bad-guy spy.

**UDE 7**
Once the blame is fixed on me,
maybe I get killed next in an
"accident".

**UDE 1**
Everyone on my team is
dead except me.

**UDE 2**
Money has been transferred
somehow to my relatives.

**UDE 9**
Whoever really killed everybody
and set me up is still out there.

**Figure 1.12. Ethan Hunt's Initial Current-Reality Logic Tree**

valid over a very great range of circumstances. So he is not surprised to find that Figure 1.12 shows that only one or a few (one) of the entities on the CRT are root causes or core problems.

He wants to move into action, to make a plan. What to do? From the diagram, and at first glance, it appears that working on anything other than UDE 9 will only prolong the problem and, most likely, make it worse.

So Ethan knows he is in trouble. He knows he will eventually need a miracle to get out of this situation, but — for the moment — he will settle for even a rough draft of a TOC future-reality logic tree, a vision statement. He lets his intuition begin to work for him.

## In Better Shape Now

It is interesting that his intuition now *will* begin to work for him. When he started to write statements down and began to think in terms of stating the facts of the situation with precision, he found that the part of his mind that was just spinning around with a lot of unproductive thoughts receded dramatically. There's another part of him dominating now, another one that he also calls his mind. He is still full of energy but is somehow more calm, focused, and deliberate. Even his body feels more a part of the action now that he has shifted from just thinking (in the sense of just having a lot of thoughts) over to *thinking* (in the sense of actively matching words and cause-and-effect relationships to his experience of what's been happening). In other words, Ethan has used the process of verbalizing his intuition to:

1.   Get started on analyzing a problem.
2.   Get his mind and emotions back into a balance, which will allow him to be effective in both thinking and, eventually, action.

Even if the current-reality logic tree he built turns out to be wrong,[46] he is in much better shape to work on some other part of the problem with clarity and intensity. If he determines that there *is* a mistake on his current tree, he will come back and fix it. This is better than being stuck and better than having already gone off in "ready-fire-aim" mode with a wrong solution. At this point — ready or not — Ethan is going to start building a game plan, a TOC future tree. Let's take a few moments to understand what a future tree is.

## Future-Reality Tree (FRT) Process and Diagram

A future-reality logic tree literally gives a picture of — call it what you will — the strategy, the vision, the mission, or the game plan for an organization or individual. It is simply a diagram that shows the ideas, the *injections*, that will be established in the situation and the desirable effects that introducing those injections will cause.

The future-tree process is used to create the future-tree diagram. The injections are formulated and written on the diagram. The individual cause-and-effect relationships, the links between the injections and both the intermediate effects and ultimate desirable effects, are established and shown on the diagram. Other facts, entities, needed to complete the picture of the cause-and-effect logic of the plan are placed on the diagram. The people who prepare the diagram, other people within the organization, and selected experts from outside the organization use the TOC categories of legitimate reservations (CLR) to test the logic of the plan. They do this entity by entity, link by link, until the logic is no longer loose or all wet, but rather as tight or dry or solid as well-cured concrete or glue.

The simple fact that the future tree is a diagram, or picture, makes it a powerful psychological and motivational influence, not only during the

## Future-Reality Logic Tree

- Injections (Ideas)

- Intermediate Effects

- Desirable Effects (DE)

- "Trunk" of Tree

- Negative Branches

- "Tight" or "Loose"

- Strategies

- Vision Statements

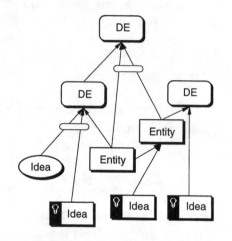

**Figure 1.13. Future-Reality Tree (FRT) Management Process and Diagram**

process of preparing it, but also throughout implementation of the plan. Figure 1.13 is a simple and generic future-tree diagram.

There are benefits to confidence, well-being, and effectiveness that come from any technique for "getting one's act together" (i.e., for approaching a state in which the individual or group understands, and is committed to, the overall plan). The TOC future-tree process delivers those benefits and more. Let's kill two birds with one stone by selecting a "sample" future tree that will also serve as your first draft of a vision statement for introducing the TOC management system into your organization or life.

## Sample TOC Future Tree: TOC Management System

Figure 1.1, the summary of the TOC management system discussed in the last section, is an example of a TOC future-reality logic tree (FRT) diagram. It links the new conditions (the ten shaded entities representing major elements of the TOC management system) to be injected, or established, into a situation with the desirable effects (in this case, strategic and operational effectiveness) that will be created when those injections have been put into place. This discussion will introduce several of the basic parameters, formats, and conventions typically used on TOC cause-and-effect logic-tree diagrams.

## Generic vs. Situation-Specific Logic Trees

This particular future tree is a generic (vs. implementation-specific) future tree in that it displays the typical cause-and-effect relationships that exist between the conditions to be introduced into the company (in this case, the elements of the TOC management system) and the desired outcomes (in this case, strategic and operational effectiveness).

## Logic-Tree Format

This tree uses the most popular format of a future-reality logic tree — placing the injections (the ideas or conditions to be established to change the system) mostly at the bottom of the page, with intermediate effects and other fact entities located primarily at the middle of the page. The sought-after results, the desirable effects, are located mostly at the top of page.

## Level of Aggregation of Logic Trees

This future-reality logic tree is displayed at a high level of aggregation in that, for example, entity 18a — Supply chain, enterprise, and critical chain schedule-based decision-support (SBDS) are in use — can (and normally does) open up into dozens of future tree pages. This is because many separate solutions (e.g., drum-buffer-rope scheduling and buffer management shop control) are aggregated into that one entity for purposes of summary and display. This level of aggregation of the future-reality logic tree allows the entire picture to be displayed on a single page.

## Viewpoint of Logic Trees

In live TOC implementations, logic trees are prepared by and for each major viewpoint (oftentimes the various departments) of the organization before they are merged into a single logic tree that shows the primary inter-relationships. Often, in complicated situations, after a full analysis is done, a communication logic tree will be formed which presents a subset of the information from the analysis in a way that "nets it out" and causes progress to be made with an important individual or group. Figure 1.1 takes the viewpoint of someone unfamiliar with the TOC management system and links the major components to the major desirable effects.

## The Example

The primary desirable effect (DE) of the future-reality logic tree is displayed at the top of the page within the entity box labeled "1a. Enterprise ... serves its targeted customers and goal". This box is entity 1a on the tree. The major elements of the TOC management system are shown within the ten highlighted and shaded boxes as entities 5–7a, 11, and 13–18a. For example, entity 7a is "7a. TOC logic trees (TP) are among the company's standard tools for verbalizing intuition".

On TOC cause-and-effect diagrams, an entity is a single fact or characteristic of the situation which will be examined in relationship to other such factors. The relationships between entities on logic-tree diagrams are established as cause-and-effect linkages, with the entity at the base of the arrow designated as a cause of the effect entity shown at the tip of the arrow.

These are "if ... then" statements. For example, consider entities 5 and 4 on Figure 1.1. The entities are to be read from the base of the connecting arrow to the arrow's tip as follows: "*If* TOC's 'intrinsic order' principle is adopted, *then* a 'can-do' attitude exists."

Another example, using entities 7a, 8, 18a, and 19: "*If* (7a) TOC logic trees (TP) are among standard tools for verbalizing intuition, and *if* (8) people work effectively on the right tasks, and *if* (18a) supply chain, enterprise, and critical chain schedule-based decision support (SBDS) are in use, *then* (19) business process designs are effective."

One final point: The numbers are assigned to the tree entities at random. They are not intended to indicate a sequence. In Figure 1.1, entity 1a is simply a revised version of some original entity 1. The purpose of these entity labels is to give them a simple name — any simple name — for use during discussions. For example, consider a hypothetical discussion:

**Joe:** "Audrey, it seems to me that 5 can cause 4 by itself, without the additional causes represented by 12 and 8."

**Audrey:** "That's what the diagram says, Joe. When there's no linking symbol across the cause-and-effect arrows, it means each cause is itself sufficient to cause the effect. The tree already says that any one of 5, 8, or 12 is sufficient to create a 'can-do' attitude within a company."

**Joe:** "I knew that."

**Audrey:** "Sure ..."[47]

With that understanding of TOC future-reality logic trees, let's find out how Ethan Hunt is doing with his planning.

## Ethan's First Future-Reality Tree (Plan) Treats Only Symptoms

The idea to go into hiding flashes into his mind. He writes it down as the first idea or injection on his future-reality logic tree.

He knows he may eventually wish to get the practical benefits of having a technically correct future tree built and tested along the lines of the rules for trees. However, right now, he is just interested in getting started and in maintaining the momentum of the problem-solving process. He once read Tom McMullen's book, *Introduction to the Theory of Constraints (TOC) Management System*. He remembered Tom's advice that the trees are supposed to serve you, and not vice versa. Therefore, it's often more important to get the benefit of the facts — and to verbalize your intuition about the facts — than it is to slow the process for considerations of the rules about trees. Ethan wasted no time in drawing the simple future-reality tree trunk detector form shown as Figure 1.14.

He then proceeded to identify and write down the pros and cons of the ideas in the boxes. The pros could include: might not die, won't go to jail, maybe more; cons: relatives still in trouble, might die anyway if the bad guy finds me, might go to jail anyway if they catch me, can't work. These are shown in Figure 1.15.

He now knows the idea he was considering injecting into his situation – going into hiding — does not lead to a solid trunk of a TOC future tree. In other words, it does not lead to strong desirable effects.

If the injection *had* led to very strong desirable effects — for instance, if the pros had been "will not die" and "will not go to jail" instead of "*might* not" — this might create the planning condition of a future tree with a strong trunk. In that case, Ethan would normally begin a sub-process called *negative branch processing*. This process would refine the thinking in the plan and/or add additional ideas or injections. These activities would make the plan more practical, more complete, and — frankly — more elegant. He could use negative branch processing to generate ideas that would increase the likelihood of staying alive and out of jail, but he knows that, when he runs out of ideas, he can turn to another tool, the ...

## Evaporating-Cloud (Conflict) Logic-Tree Process and Diagram

### Fresh Out of Ideas? Break Out the Evaporating-Cloud Process

The primary role of the evaporating-cloud process is to generate new ideas for the planning process. It does this by providing a systematic way to

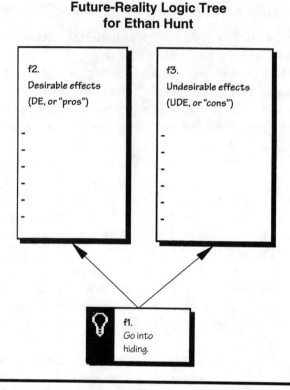

**Figure 1.14. Ethan Hunt's Draft of a Future-Reality Tree Trunk Detector**

verbalize the explicit and implicit assumptions made by people involved in the situation. Once articulated, the assumptions can be challenged in order to identify ideas for the needed plans. Another role for the cloud diagram is to communicate an aspect of the situation for purposes of building consensus on the plans.

There are several formats for the evaporating-cloud diagram. The two primary ones are the objective-requirements-prerequisites format and the objective-needs-wants format. The latter format is shown in Figure 1.16.

## Example of Needs-and-Wants Cloud Format

For purposes of illustration, we will assume that the situation depicted in Figure 1.16 is that two parties, Clara and Will (who, by the way, have nothing to do with the *Mission Impossible* story) share the same objective, entity 1. This objective could be something such as "have fun together." (Suggestion

## Future-Reality Logic Tree
## for Ethan Hunt

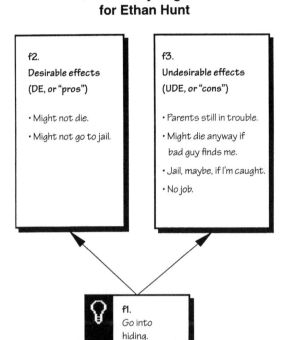

**f2.**

Desirable effects

(DE, or "pros")

• Might not die.

• Might not go to jail.

**f3.**

Undesirable effects

(UDE, or "cons")

• Parents still in trouble.

• Might die anyway if bad guy finds me.

• Jail, maybe, if I'm caught.

• No job.

**f1.**
Go into hiding.

**Figure 1.15 Ethan Hunt's Second-Draft Future Tree: No Solid Trunk of a Future-Reality Tree Is Indicated; It Treats Only Symptoms, Delivers Less than a Cure, and Delivers Even Those Results With a Low Degree of Certainty ... It's Back to the Drawing Board for Ethan.**

for the reader: As an exercise, draw the five boxes of an evaporating-cloud logic-tree diagram and fill them in as you read this.)

Entities 4 and 2 are the *want* and *need,* respectively, of one party (Clara). Entities 5 and 3 are the *want* and *need* of the other party (Will). Both sets of wants and needs support the common objective. The conflict is indicated by the double-headed arrow between entities 4 and 5. For example, the friends cannot both go to the movies now (Will's want, entity 5) and go to the shore now (Clara's want, entity 4).

Will has cabin fever; his need (box 3) is to get out of the house. Clara, at her house, is convinced she needs to work on her tan. She is aware of all the ozone, ultraviolet, and infrared issues, but — hey! — the lady likes to have a tan. Will doesn't want to hold *that* against her.[48] Anyway, they know they have this conflict and agree to draw the cloud diagram together.

## Evaporating Cloud (EC)

- Conflict process and diagram

- Objective

- Needs

- Prerequisites

- Inherent conflict

- Examines assumptions

- Generates ideas

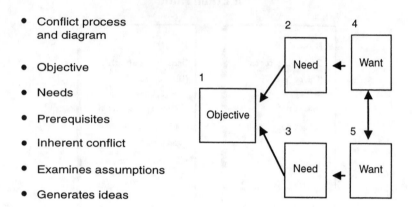

**Figure 1.16. Evaporating-Cloud (Conflict) Logic-Tree Management Process and Diagram**

Will reads his part: "The objective we share is to have a good time together today. That's entity 1. In order for me to have a good time today, I *need* to get out of the house (entity 3). In order to do that, I really *want* to go to the movies (entity 5). Hmmm … Wait a minute." He looks at Clara and says, "Well, I guess I'm making an assumption on the arrow from entity 5 to entity 3 that, in order to get out of the house, we have to go to the movies. That's not a valid assumption."

Clara says, "Right, and I guess I've been making an assumption on the arrow from 4 to 2. I've been assuming that, in order for me to get the very much larger tan I *need* (now that you so generously have bought me a new swimsuit), that I must go to the shore to get it."

Unaccountably, Will suddenly becomes very enthusiastic. He proposes an idea: "Why don't we just catch some rays in the backyard?"[49] Clara sees the potential immediately and shares his enthusiasm.

The idea has removed the conflict. It has removed the problem. It has evaporated the cloud. This new idea is an injection suitable for placement at the bottom of a TOC future tree. They decide not to draw the new future tree, at least for this case, because (1) it is a simple case, (2) they both see the benefits clearly, and (3) time's a-wasting.

Clara suits up to get her new tan. She's happy. Will … well, Will sort of lost his interest in whatever the name of that movie was that he thought he wanted to see so badly. As Clara said, he's a very generous guy. Anyway, that's the needs-and-wants format of Figure 1.16: quite a figure!

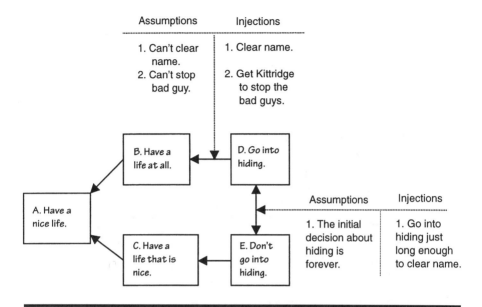

**Figure 1.17. Diagram of Ethan Hunt's TOC Evaporating-Cloud Tree (ECT)**

## Other Formats for Evaporating-Cloud Logic Trees

The other primary format replaces the needs with requirements and the wants with prerequisites. When considered in a more general way, the needs and requirements (entities 2 and 3) can be thought of as higher level requirements, while entities 4 and 5 can be thought of as lower level requirements.

## Ethan Uses the Objective-Requirements-Prerequisites Format

Let's get back to the action with Ethan Hunt. Figure 1.17 is an evaporating cloud diagram of the objective-requirement-prerequisite type. It is also of the "do it" vs. "don't do it" type of cloud. The dilemma that arose when he started to build and test his plan on a future-reality logic tree was "go into hiding" vs. "don't go into hiding". Using Figure 1.17, he identifies several assumptions. Three of the many assumptions — the two on the arrow between B and D ( B ← D) and the one on the conflict arrow between D and E (D ↔ E) — give rise to ideas, injections, right away.

Unlike Clara and Will, Ethan will use these injections as inputs for his new future tree. While Ethan works on that diagram, we will point out that the evaporating-cloud logic-tree management process and diagram are sometimes also known as:

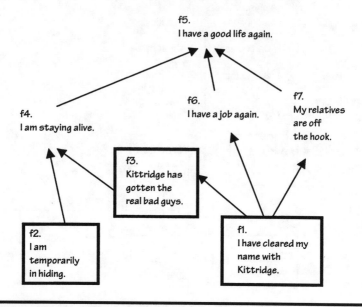

**Figure 1.18. Ethan Hunt's New TOC Future-Reality Logic Tree**

- Conflict-resolution diagram
- Conflict logic tree
- Conflict tree
- Dilemma logic tree
- Dilemma tree[50]

The name matters less than the process that offers simplicity, power, and a very wide range of applicability.

## Injections from Cloud Process Form New Future Tree

Figure 1.18 is Ethan Hunt's new TOC future tree. The undesirable effects from the TOC current tree in Figure 1.12 are all converted by this plan into desirable effects.

## Could Have Gotten There More Quickly

As Ethan considers his new future tree, he whacks himself on the forehead because he is reminded all over again that he is in Stage Three of TOC logic-tree skills, not Stage Four. And how does he know this? Because he has wasted some valuable time in getting to the sound vision, or strategy, statement of Figure 1.18.

## Failed to Apply TOC's CLR[51]

Here's what happened. He tried to save some time when building the current-reality logic tree by skipping steps, and by skipping some of the categories of legitimate reservation (CLR). For example, if he had applied the test of the CLR's causality existence reservation to the links on Figure 1.12 from UDE 9 to UDE 1 and UDE 2, he would have found that they are, well, hogwash.[52] The fact that the *real* bad guys are still out there is not the cause of everyone but Ethan being dead, and of the money having been transferred to his relatives' bank. The *real* bad guy did those things, but that he is still out there is not the current cause of those things that happened in the past. Fine distinction maybe, but the difference between getting huge results from the least effort and getting weak or bad results from huge efforts almost always turns on subtle distinctions in facts, objectives, and plans. Hold that pose. Let's state that one again, because these TOC logic-tree disciplines are not designed to turn everyone into nitpickers; they are intended to focus resources for huge effect. So here it is again.

> The difference between getting huge results from the least effort and getting weak or bad results from huge efforts almost always turns on subtle distinctions in facts, objectives, and plans.

There it is. Now, *you* have to make sure that the refinement you are making during a TOC planning session is in the spirit and flow of identifying the breakthrough solution and not just nitpicking. How? Let your intuition and your integrity be your guide. May the force be with you.

## Categories of Legitimate Reservations

The TOC categories of legitimate reservations (CLR) are clarity, entity existence, causality existence, cause insufficiency, additional cause, predicted effect existence, and tautology. TOC thinking-process expert, Bill Dettmer, likes to add cause-effect reversal to the list.

Just take a moment to think about these categories. Their titles alone say a lot. Their simple and systematic use delivers more. You can start using them right away by figuring out what they mean from their names, or you can consult logic-tree reference sources to read pretty much what you've already figured out yourself from just thinking about the names.[53]

As time, competition, and marketing progress, there will be competing logic-tree experts, logic-tree schools of thought, and several versions of these legitimate reservations. The situation will be a little like playing the card game Rummy. Hoyle and other experts have their own sets of specific rules that go with specific names of the games. There is a lot of overlap among the experts' versions and a few differences. Some differences matter. Other differences are some wanna-be expert putting his or her own spin on things just to create the perception of a niche. In practice, in order to play the game at all, people will have to decide which recommended rules they're going to use, which creates a sort of "house rules" for the game. Some people like to do this.[54]

If you do this — if you create a set of "house rules" at your company — be sure they accomplish the purpose of the CLR, which is to remove as much as possible of the non-constructive effect of personality, politics, and power from the critical task of identifying the intrinsic order that exists within the situation at hand. If Mother Nature has provided a breakthrough opportunity for your company, you must work very hard to ensure that Human Nature doesn't force you to leave it on the table.

## Encourage Study of the TOC Logic-Tree Rules

By creating a loose (vs. tight)[55] current-reality logic tree, Ethan arrived at a less correct expression of the core problem, the entity that, if changed, would cause the undesirable effects of the current tree to go away. If UDE 9 on Figure 1.12 were removed by, say, Ethan's finding and removing the bad guys, it would only help with UDE 7 (the risk he'll be killed by the real bad guys). Notice also that, had Ethan followed the TOC logic-tree disciplines which have been worked out, he would also have shown a cause-and-effect arrow from UDE 3 to UDE 7. This is because removing the real bad guys only removes *one* of Ethan's most serious and immediate risks of getting killed. Without also clearing his name, Ethan is also at high risk of being killed by government intelligence agents.

What he is seeing in his new future tree — which was developed from the verbalization of intuition, the surfacing of assumptions, and the generation of ideas or injections, all via the evaporating-cloud process — is that he had identified the wrong core problem. This is why he wasted time spinning his wheels on the fruitless initial future trees.[56] Too bad. It is important to realize that going from Stage Three to Stage Four in TOC logic-tree skills is worthwhile.[57]

## But, For Heaven's Sake, Don't Tie the Do-ers Down!

On the other hand, it is also crucial to notice that Ethan got to the right place even though he did not dot all the *i*'s and cross all the *t*'s in exactly the right sequence, at every step along the way.

This is a crucial point, in order to keep the inevitable class of TOC logic-tree process geeks[58] from blocking the energy, momentum, contribution, and effectiveness of the real resource of any organization, the powerful and intuitive action folk who brilliantly and intuitively balance the requirements and benefits of analysis and action.

Let me say this in another way: If Ethan's intuition were so strong that he was able to will himself to create the right TOC future tree on the spot — with no current tree, no cloud, no intermediate steps — that would be great. The best and most effective people of action can do that, especially in emergencies. (Not so fast, you brilliant intuitives. You are not off the hook yet. The TOC logic trees are excellent tools for you to communicate better, to yourself and others, about the underlying logic of your strokes of genius. And there *is* a logic to each such stroke.)

There are data in the world and based on experience that say people are able to do that. Since TOC is a science, it acknowledges those facts along with all the others. So TOC is not designed to get in the way of that. Quite the contrary, TOC would work at sorting out the cause-and-effect relationships involved in giving rise to greater numbers of people who can and do rise to occasions.

At the same time, most situations and most individuals and most groups should have a deliberate process for examining the thinking in a situation and for systematically converting huge and complicated problems into huge and breakthrough solutions.

An important purpose of the TOC logic-tree thinking processes is to further empower do-ers – vigorous, practical, visionary, dynamic, and effective action people. The purpose is emphatically *not* to give the nervous nellies[59] who live in every organization of any size (i.e., two or more people) yet another cotton-picking, nitpicking, spirit-wasting tool to strangle the life out of the type of vigorous people who are the life of institutions, especially during times of crisis.

## Learn True TOC Logic-Tree Process From the Do-ers

When your best and most effective people tell you that the TOC logic trees are "slowing them down" rather than "slowing them down only temporarily in order ultimately to speed them up and ultimately dramatically increase the magnitude of their constructive impact", then listen to them and learn from them about how you and your company *should* be using the TOC logic trees. If your people are already strong and effective, their intuition is healthy. Even if the Ethan Hunts in every organization are doing it "wrong" for less important measures (such as the nits and nats of the exact baseline logic-tree process steps) but "right" in terms of getting to the results, don't get in their way and don't tie them up. Their instinct for more results, more results more quickly, and more results more quickly with ever fewer resources will fine-tune their own and the entire organization's techniques for using TOC logic trees. They will find the ways to use the tools that work for them and integrate the fundamentals of the logic trees into the culture, language, and process of your institution.

What's that? Oh, right. Sorry. I guess I got a little carried there away, didn't I? This topic of Ethan — who is a brilliant and terribly effective individual — getting it wrong but then getting it right, and what that means about how to deal with your various types of people as they learn to use TOC, is, well, sort of important. But I guess I have pretty much made that point clear once or twice already, right? Okay. So, we will talk about something else for awhile. (Say, how about those Red Sox!) Let's get back to the action with Ethan.

## Ethan Returns to and Corrects the Current-Reality Tree

Figure 1.19 is Ethan's revised current-reality logic tree.

## Prerequisites Tree (PRT) Process and Diagram

So far, so good. However, Ethan now has a new problem. It is fine to say that the plan — the strategy, the vision — expressed on the future-reality logic tree of Figure 1.18 is sound. It is quite another thing to know how some of the more challenging aspects of the plan are going to be accomplished.

It is clear that the f1 injection, "I have cleared my name with Kittridge", is the correct condition to establish. It is also clear that accomplishing this will be no trivial project.

This is the role of the TOC prerequisites tree (PRT), to break an important objective, such as a crucial injection from a TOC future-reality logic tree, down into practical intermediate objectives (IO). It's like eating an elephant — a bite at a time! Figure 1.20 shows the generic structure of a TOC prerequisites logic-tree diagram. The process for creating a prerequisites tree is straightforward.

1. *Establish the objective.* We already know the objective: Clear Ethan's name. As with the future-reality tree injection, the objective is stated and written as if it has already been accomplished. This provides a subtle, but important, motivational boost, especially when what you might call "impossible missions" must be undertaken. The objective is written as "My name has been cleared with Kittridge". This PRT entity is placed at the top of the diagram.

2. *Identify the obstacles to reaching the objective as well as enough intermediate objectives (IOs) to overcome all the obstacles.* Ethan knows this

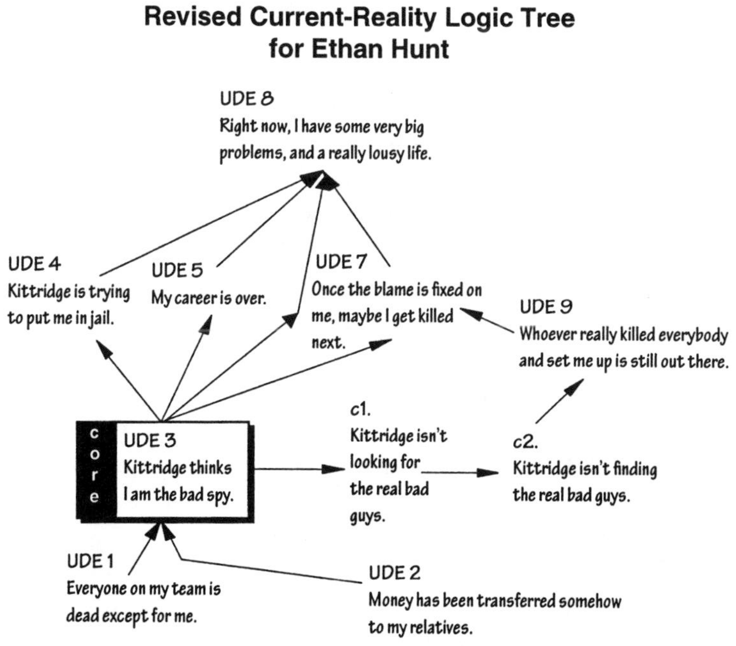

**Revised Current-Reality Logic Tree for Ethan Hunt**

**UDE 8**
Right now, I have some very big problems, and a really lousy life.

**UDE 4**
Kittridge is trying to put me in jail.

**UDE 5**
My career is over.

**UDE 7**
Once the blame is fixed on me, maybe I get killed next.

**UDE 9**
Whoever really killed everybody and set me up is still out there.

**core UDE 3**
Kittridge thinks I am the bad spy.

c1.
Kittridge isn't looking for the real bad guys.

c2.
Kittridge isn't finding the real bad guys.

**UDE 1**
Everyone on my team is dead except for me.

**UDE 2**
Money has been transferred somehow to my relatives.

Figure 1.19. Ethan Hunt's New TOC Current-Reality Logic Tree

**Prerequisites Tree (PRT)**

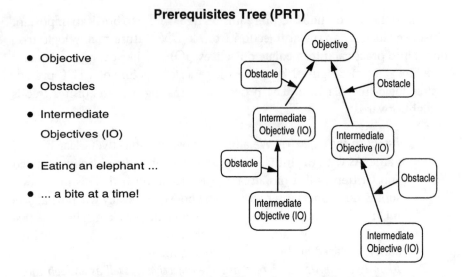

- Objective

- Obstacles

- Intermediate
  Objectives (IO)

- Eating an elephant ...

- ... a bite at a time!

**Figure 1.20. Prerequisites Tree (PRT) Management Process and Diagram**

procedure. He's done this before. In fact, he likes this part because it comes so naturally. All he has to do is make believe he's a whining and nay-saying type (he isn't) and make a list of the reasons it will never work. He creates the table shown in Figure 1.21.

3. *Arrange the intermediate objectives (IO) and obstacles into the structure and sequence of the prerequisites tree.* Ethan needs to work out the sequence of intermediate objectives that will accomplish the objective. Ethan Hunt's TOC prerequisites logic tree is shown as Figure 1.22.

As it happens, Ethan will need at least three prerequisites trees. First, the complexity of the operation required to get help from other ex-agents disavowed by the Impossible Mission (IM) organization and then obtaining the real list of deep-cover spies from the IM headquarters in Langley merits its own planning process and the first TOC prerequisites tree. This PRT he will share with Claire Phelps, Luther, and the knife guy because they will be cooperating to steal the computer disk of names. Second and third will be two different prerequisites trees for the London-to-Paris chunnel (channel tunnel) train operation. The second PRT will show the plan that Ethan wants Jim and Claire Phelps and the knife guy to *think* is Ethan's plan to get money from

| Obstacles | Intermediate Objectives (IO) |
|---|---|
| I don't know who the bad guy is yet. | Go to the bad guy's customer |
| I don't have a way to find the bad guy | Ditto. |
| Kittridge will try to capture me. | Don't let him get close enough (don't stay on the phone too long with him, use eyeglass video transmitters, etc.) |
| I don't have a way to make the bad guys come to me … wait a minute … | *Injection* — The bad guys thought they got the real spy list, and they will still want it. *Injection* — I can get the list and have them come to me. |
| Having them come to me doesn't help me with clearing my name. | Have the bad guys come to Kittridge. |
| I don't have a way to have the bad guys come to Kittridge. | Get the spy list, and have both Kittridge and the bad guys come to me at the same place. |
| I don't have a way to get Kittridge to come to me. | Kittridge will come to me if I tell him where I am. *Injection* — Call Kittridge. Stay on phone long enough to trace me to London, but not where in London. |
| No way to show real bad guy to Kittridge. | *Injection* — Use the eyeglass-mounted videocameras to project picture of bad guy when they come to me to wristwatch. |
| No way to get wristwatch on Kittridge and within transmission range of eyeglass video transmitter. | Do several things: 1. Give wristwatch to local spy office in a package addressed to Kittridge; they will surely give package to Kittridge. 2. Limit maximum range by making the event happen on a train. 3. Limit ability to be tricked or set up by selecting very fast train. 4. Tell Kittridge about train by supplying tickets in package with wristwatch video receiver. 5. Plan to wear video transmitter eyeglasses on train, so when real bad guy and I meet, Kittridge will see and hear the whole thing. |
| I don't have a way to contact the bad guys. | *Injection* — I know they have been communicating in the Job section of the Internet. |

**Figure 1.21. Ethan Hunt's List of Obstacles and Intermediate Objectives for a TOC Prerequisites Tree (PRT)**

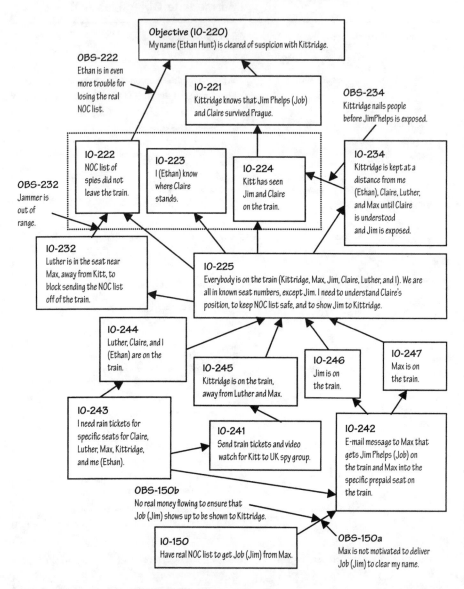

**Ethan's Prerequisites Tree (PRT)**

**Figure 1.22. Early Draft of One of Ethan Hunt's TOC Prerequisites Trees Used To Clear His Name and Get His Life Back; Two Other Full PRTs Were for (1) the Langley Operation and (2) the Fake London-to-Paris Plan To Confuse and Set Up Claire, Jim, and the Knife Guy**

Max, the arms merchant, to get a helicopter escape ride from the knife guy upon reaching Paris and then to run away happily ever-after in love with the money and Claire Phelps. However, since he is pretty sure the knife guy (Krieger) is a bad guy working with likely bad guy Jim Phelps and since he is not yet sure about Claire, he has a third prerequisites tree that is the *real* plan — the trap on board the chunnel train — to figure out where Claire stands, allow Kittridge to see and apprehend a very much still-alive Jim Phelps, and, in the process, clear Ethan's name and get his life back in order. Figure 1.22 is that third prerequisites tree.

## Touring the Sample Prerequisites Logic Tree

A TOC prerequisites logic-tree diagram begins with the objective. This is the entity at the top of Figure 1.22 (IO-220): "My name is cleared of suspicion with Kittridge." This is also one of the three injections on Ethan's TOC future tree shown in Figure 1.18. Getting PRT objectives from future-tree injections is one of the two most typical ways of establishing the objective. The other is simply to write the objective that is known, even if it was not developed through use of the future-reality tree process.

At the other end of Ethan's PRT is entity IO-150: "I have the real NOC (spy) list for use in getting Max to help me get Job (Jim Phelps)." This entity and outcome are the objective and the result of the prerequisites tree used to plan and implement the successful operation at Langley. The result is that Ethan and his team now have the *real* list of spies, which allows him to use Max's influence with Job (Jim Phelps) and Claire's money as the bait for the trap that will expose Phelps and clear Ethan's name.

The relationship between the PRT's *intermediate objective* entities and its *obstacle* entities is also shown in Figure 1.22. To establish the circumstance of IO-150 is to anticipate and preemptively overcome the two obstacles — PRT entities OBS-150a and OBS-150b — that would otherwise block Ethan's ability to get to Job/Jim via Max's cooperation. On a TOC PRT diagram, the intermediate objectives are connected with an arrow between each two IOs. The obstacles that are overcome by the IOs are shown with arrows that lead from the obstacle entity to the arrow between the two connected IOs.

Another example of an intermediate objective being planned in order to anticipate and overcome predicted obstacles is IO-234 at the upper right portion of Figure 1.22. One of the potential obstacles (entity OBS-234) to Ethan clearing his name is that Kittridge might recognize and apprehend him, his teammates, or Max before Kittridge had seen Jim Phelps alive. This would prevent Ethan from carrying out his plan to trap Jim Phelps. This potential obstacle is anticipated and preemptively overcome by establishing the condition indicated in entity IO-234. This involves maintaining a physical separation between Kittridge and the other key players until Kittridge has seen Phelps on the wristwatch video.

Figure 1.22 is a picture of where Ethan left off on his PRT before moving on to another TOC tool, the transition logic tree (TT). The process of making the chart in Figure 1.21 was helpful both in keeping Ethan in good emotional shape and in making progress toward the desired condition. The process of making the good, but not perfect, PRT in Figure 1.22 clarified many details of necessary sequence. It also highlighted that even more detail would be required in order to determine how to get from IO-225 ("All the right people on the train") to IOs-222, 223, and 224 ("Phelp's seen by Kittridge", "Claire's agenda is known", and "NOC list of spies is not compromised"). Events will move quickly on the chunnel train. More meticulous detail is needed to ensure that the encounter accomplishes all the important intermediate objectives. This is the purpose of the transition tree, which will be discussed in the next section.

Ethan's prerequisites tree for the trap at the chunnel train was not complete from a strict technical point of view. For instance, it lacked obstacles between all IOs and did not have all dependencies characterized with precision. Still, it was *reasonably* complete, had revealed many important issues, and had clarified several critical matters of sequence. His instincts were now telling him that improving the prerequisites tree from its current state of "good enough for now" was less important than shifting focus immediately to establish some appropriate level of detail in the transition-tree planning diagram for the critical chunnel train operation.

The situation was certainly important to him, so — if he had time — he would take time to fine tune the full set of trees. He would do this in order to increase the chances that he would achieve the maximum progress toward his objective, with minimum risk, and with minimum consumption of resources.[60] In fact, when Ethan has time, he usually has internal and external expert technical, legal, and intelligence specialists review the appropriate entities and cause-and-effect links in his logic-tree diagrams.

## *Transition Logic-Tree Process and Diagram*

Transition logic trees, also known as transition trees, can be used to:

## Transition Tree (TRT)

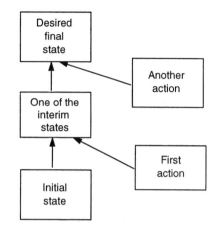

- Need for action

- Objective

- Sequence of states

- Sequence of actions

- Sufficiency

- One-time changes

- Recurring processes

**Figure 1.23. TOC Transition Tree (TT) Management Process and Diagram**

■ Document existing and new recurring processes (e.g., sales processes, forecasting processes, scheduling processes, new product development processes, Little League baseball at-bat processes, screenplays)

■ Accomplish one-time changes in circumstances (e.g., establishing new facilities, improving business processes, making strategic moves, and, of course, getting out of really complicated and dangerous situations — such as those in the spooky world of our hero, Ethan Hunt)

As shown in the simplified generic transition tree in Figure 1.23, the transition tree is a series of action entities and state entities. The actions create the changes in state, the state of some situation. Not shown in Figure 1.23 are three other elements of the most comprehensive format for the transition tree's repeating element. These very useful entities have to do with demonstrating needs at successive stages of the managed transition and with demonstrating that the action taken at each stage will be sufficient to accomplish the projected change in state. The prerequisites tree is only called for if the future tree's injection is sufficiently complex to justify the additional planning and detail. Similarly, a transition tree is not prepared to support one or more prerequisites-tree intermediate objectives unless their difficulty or importance merit it. Similarly, the two *needs demonstration entities* and the *single sufficiency demonstration entity* of each transition-tree cycle are only used when the benefits of the incremental time and trouble so warrant.

## Logic-Tree Tools for Successive Levels of Detail

Ethan's TOC prerequisites tree shows the intermediate objectives (the stepping stones) along the path to clearing his name with Kittridge. Just as the prerequisites-tree process "unbundled" a complicated and challenging injection from the future tree by breaking that objective down into smaller pieces, so, too, does the TOC transition tree provide additional detail in planning for selected and more difficult intermediate objectives (IO).

## One Intermediate Objective Is Tricky and Important Enough To Warrant the Additional Detail of a Transition Tree

One IO that is very important to Ethan — and will be tricky — is ensuring that Kittridge will get enough of a show on the eyeglasses-to-wristwatch video channel to clear Ethan's name of suspicion. This will take some doing. By the time the confrontation on the speeding train outside London occurs, Ethan is fairly certain that Jim Phelps is the bad spy who sold the list of names to Max, the arms merchant, and killed everyone else on the team. He knows Claire will be on the train, because she is working with him. He does not yet know if she knows Jim Phelps is still alive or if she is part of the dirty dealings.

In other words, when the London-to-Paris chunnel train part of the plan begins, Ethan will be facing a condition or a state of needing to:

1. Know whether Claire Phelps[61] knows Jim Phelps is still alive
2. Know where Claire stands on the murders and frame-up
3. Show Kittridge that other members of the team are still alive
4. Prevent the *real* stolen list of spies (the NOC list from the Langley operation) from being compromised (or "getting into the open")

This is the entity shown as "455 Need" on Figure 1.24. This "need statement" entity supports transition step number 455. The reason the number (455) is so high is that many transition tree steps were needed for the operations to find and get support of disavowed IM agents, to conduct the NOC list operation at Langley, and for operations too detailed or sensitive to show in the movie. The entity labeled "455 State" details the objective that step 455 is intended to accomplish. Entity "455 Actions" describes what will be done, while entity "455 Sufficiency" states why, given the initial conditions summarized in the need statement, that the actions will be sufficient to reach the desired new state of circumstances. In this situation, adequate precision and

## Ethan's Transition Tree (TRT)

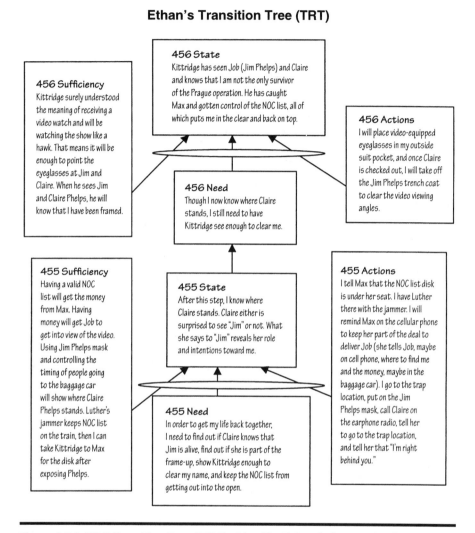

**456 State**
Kittridge has seen Job (Jim Phelps) and Claire and knows that I am not the only survivor of the Prague operation. He has caught Max and gotten control of the NOC list, all of which puts me in the clear and back on top.

**456 Sufficiency**
Kittridge surely understood the meaning of receiving a video watch and will be watching the show like a hawk. That means it will be enough to point the eyeglasses at Jim and Claire. When he sees Jim and Claire Phelps, he will know that I have been framed.

**456 Need**
Though I now know where Claire stands, I still need to have Kittridge see enough to clear me.

**456 Actions**
I will place video-equipped eyeglasses in my outside suit pocket, and once Claire is checked out, I will take off the Jim Phelps trench coat to clear the video viewing angles.

**455 Sufficiency**
Having a valid NOC list will get the money from Max. Having money will get Job to get into view of the video. Using Jim Phelps mask and controlling the timing of people going to the baggage car will show where Claire Phelps stands. Luther's jammer keeps NOC list on the train, then I can take Kittridge to Max for the disk after exposing Phelps.

**455 State**
After this step, I know where Claire stands. Claire either is surprised to see "Jim" or not. What she says to "Jim" reveals her role and intentions toward me.

**455 Need**
In order to get my life back together, I need to find out if Claire knows that Jim is alive, find out if she is part of the frame-up, show Kittridge enough to clear my name, and keep the NOC list from getting out into the open.

**455 Actions**
I tell Max that the NOC list disk is under her seat. I have Luther there with the jammer. I will remind Max on the cellular phone to keep her part of the deal to deliver Job (she tells Job, maybe on cell phone, where to find me and the money, maybe in the baggage car). I go to the trap location, put on the Jim Phelps mask, call Claire on the earphone radio, tell her to go to the trap location, and tell her that "I'm right behind you."

**Figure 1.24. TOC Transition Tree (TT) To Identify Claire Phelps as Friend or Foe and To Show the True Job to Kittridge, Clearing Ethan's Name**

clarity to support the operations are achieved with the statements as written. Ethan knows that, for more complicated situations, the state entities can be detailed lists of parameters and values, or even complete TOC future-reality logic trees. In the latter case, the succession of states becomes a succession of future trees, a concise and clear way to map out a long-range series of circumstances that actions are intended to accomplish for an individual, a company, an industry, a nation, or other system. At the end of transition-tree step 455, Ethan knows where Claire stands. At the end of transition-tree step 456, Ethan is in the clear.

A comment about plans, chance, and serendipity: As it turned out in the movie, the NOC list of deep-cover spies was prevented from getting "into the open" only partly by the plan on Ethan's transition tree. The plan was for Luther, one of the disavowed IM agents Ethan had hired, to sit near Max on the chunnel train and jam the cellular phone she would use to transmit the NOC list off of the train. As it happened, Luther had to leave his seat to avoid being seen by Kittridge. A well-meaning waiter then "helped" Luther by turning off the jammer. This allowed Max to begin transmitting the list. The jamming had delayed the transmission long enough that it was blocked by the train entering the London-to-Paris tunnel under the English Channel. When Ethan and Luther were in the clear, Luther helped Kittridge retrieve the computer disk from Max which finally got the list back into secure hands. The lesson? That's the way life is. The best you can do is plan well, add margins of safety, prepare hedging plans — and then make sure luck, or fate, or destiny are on your side. How to do that? Lots of ways. Each to his or her taste. But there is one that has enjoyed wide and long-standing appeal. Psalm 23, anyone?

In the end, Ethan's meticulous planning and implementation — which made effective use of a current-reality tree, a few future trees, an evaporating-cloud logic tree, several prerequisites trees, and, finally, several transition trees — paid off by putting Ethan back in the clear, and back on top!

## Better Use of Meetings

Sure, the logic trees are useful for spies on the run, but what about for use in the day-to-day life of other kinds of individuals and organizations?

## TOC Thinking Processes: Time Is the Prime Constraint

Consider this question: How many times have you sat through meetings in which nothing, too little, or the wrong things were accomplished? How many times was it because there was no clarity regarding what was being discussed? How many times was it because one individual in the group had set up the psychological situation in such a way that no one dared question them or indicate anything but the most profound respect, make that deference?

Intel's Andy Grove, in *High Output Management,* points out that the pressure *not* to have meetings at all misses the point of the nature and purpose of managerial work. The answer is to make meetings more effective. The TOC logic-tree processes, with their categories of legitimate reservations (CLR), provide the framework, process, and rules to keep a serious discussion on track. Time is used better than ever.

## *Initial Frustration, but Worth the Trouble*

People can have troubles with the logic trees — sometimes the same troubles they have always had disciplining themselves to set aside time for real thinking. An independent study team found this situation at one company:[62]

> "Ironically, there is more frustration and dissatisfaction with the level of involvement in the Thinking Process at this company than in any other company we visited. And yet the formal Thinking Process is used more often and by more people here than anywhere else, with the exception of Kent Moore Cabinets. Logic trees are used to address all kinds of problems at all levels of the company. Indeed, the president, Charlie Van Dyke, would like to see a formal TP analysis of any major decision that comes across his desk. For example, at the time of our visit, trees were being used to develop the annual business plan. There also was ample evidence that trees are used frequently by some people at the local level, although often with the help of one of the TOC facilitators from corporate headquarters."

## *Applications of TOC Logic-Tree Processes*

Chapter 6 will contain discussions of using TOC logic trees in support of various activities, such as strategic planning, policy development, human resources, day-to-day management skills, compensation, vendor management, organization design and development, operational decisions, degree of out-sourcing, degree of vertical or horizontal integration, project management, and process improvement and reengineering.

## *Other Thinking Processes*

The logic-tree thinking processes of TOC are not the only thinking processes ever invented. They may be, however, the single most simple, practical, and yet still most comprehensive general-purpose management thinking processes ever invented. Here are a few other popular processes that came before TOC.

## Ben Franklin's Decision Process

The time-honored system from Franklin's *Autobiography* has served many people well in many situations over the generations. It is a simple listing and weighing of pros and cons. It *does* invite the operation of intuition in the weighing of the factors. It *does not* even begin to approach the power of the evaporating-cloud and future-reality tree processes for digging systematically into the underlying assumptions creating the situations and choices.

## Lastname1-Lastname2

There is a well-known business problem-solving and decision-making process that is known as the hyphenation of two people's last names. I was trained in it during one of my early corporate assignments with a Fortune 500 industrial company and once watched in horror as one otherwise very successful executive got himself fired by trying to use it in a situation beyond its range of validity. It is representative of a wide range of similar processes promoted under various names. The process asks a group to list several problems or opportunities. A vote is taken to determine the relative importance of each idea. Another vote is taken to assign a numerical score indicating each idea's likely effectiveness. And another vote is taken to determine which things will actually get done. There are many problems with this approach.

For one thing, the process does not force the participants to think together, or with precision, about the nature of either the problems or the proposed solutions. If no one in the room has their finger on the real root causes and real breakthrough solutions, or if someone who does cannot persuade the group back to and beyond the wrong directions set in the initial votes, then the group will emerge feeling the relief from tension that any decision provides but will be working on the wrong things. Believe it or not, this process is not completely bad. In an environment where competition is relatively weak, staffing is not lean, and industries are forgiving of firms seeking to recover from missteps, this process will, in many situations, be right enough to deliver *some* progress. People feel good coming out of this process (except for the few whose insight and intuition are killing them because they know the group never came anywhere near the real issues). Where this class of process fails is in environments like today's in which it *does* matter whether all or most decisions are made intelligently.

## Total Quality Management

Total Quality Management (TQM) offers a variety of thinking tools, such as the fishbone (Ishakawa) diagram, quality-function-deployment (QFD) matrices, the five whys, and a type of root-cause analysis. At the 1997 APICS Constraints Management Symposium held in April 1997 in Denver, a team of experienced quality professionals and company managers reported the findings of their study comparing typical TQM tools with the logic trees of the TOC thinking processes. They concluded that, while the TQM tools are certainly very useful in a wide range of circumstances, the TOC logic trees could be used virtually anywhere and offered significant advantages in terms of rigor, speed, clarity, integration, and communication support.[63]

## Systems Dynamics and Fifth Discipline

Consultant Peter Senge and his mentors and colleagues at the MIT school of systems dynamics have done a very nice job of articulating many cause-and-effect relationships common to many systems. The side-by-side columns of "what was said" vs. "what was really said" in the *Fifth Discipline* are helpful analytical tools. The looping structures and archetypes are also useful in certain situations. Like Stephen Covey's work (discussed in depth in Chapter 6), Senge's writing has the effect of centering the reader, which, of course, means Senge is working at the heart of matters.

The Theory of Constraints and *Fifth Discipline* approaches can be used effectively together. The *Fifth Discipline* concepts help members of complex organizations to visualize the differences between departmental and systems-wide thinking and to grasp the importance of the latter. TOC provides the comprehensive infrastructure of concepts and tools including science attitude, goal, physical and policy constraints, necessary conditions, organization focusing, relative improvement rates, global and local measurements, importance scales, structured thinking processes, project management, strategic planning process, strategic combinations, process diagnosis and design, management skills, decision-support, logistics, and scheduling tools.

## TRIZ

TRIZ is a structured approach to invention which seeks to codify physical characteristics that can be applied to inventions and to further codify the invention challenges and solutions in ways that speed creation of breakthrough solutions. For example, there are rules about the hysteresis effects in metals and the temperature effects of fluids. There is also an inventor's algebra that mathematically expresses combinations of characteristics of previous or promising inventions.[64,65]

Like TRIZ, TOC can be, and is, used to make the logic of "hard sciences" and "engineering" thinking explicit. However, TOC is also used readily over a much wider scope of situations. For example, in Chapter 3 of this book, among the cases mentioned is the $4.5 billion global military hospital organization that used TOC logic-tree thinking processes to rework its strategies and processes.

The two thinking processes seek to accomplish different things and are complementary. Companies should TOC and TRIZ together in the product- and process-development portion of their work. TOC current and future trees should be used during strategic planning processes to sort out initial specifications for what the technology organizations should try to "invent" to support the plans. The inventors should use TRIZ to invent what's asked or, where appropriate, engage the strategic planners in a communication process to adjust the strategies and specifications.

## TOC Thinking Processes and Day-to-Day Management Skills

This topic deserves not only its own section but also a book-length treatment. Not here, though. For the purposes of this introductory volume, suffice to say (1) the logic trees and other thinking processes of TOC have been demonstrated to be useful in day-to-day managing, and (2) it is a useful exercise for the reader to express — in the language of the TOC logic trees — the underlying logic, conflicts, and processes of the recurring day-to-day tasks in areas such as leadership, delegation, empowerment, motivation, communication, and teamwork.

For those with a strong interest in this area, Dr. Goldratt's lecture at the 1996 APICS Constraints Management Symposium called "Overview of TOC Applications for Industry"[66] and his article in the April 1997 issue of *APICS — The Performance Advantage* magazine titled "A TOC Approach to Organizational Empowerment"[67] contain examples to get you started in your thinking.

# TOC Project Management

## Overview of TOC and Project Management

### General

TOC's logic-tree processes keep projects on track.

### Strategy

Projects that create an organization's strategic positioning are the most important and powerful applications of TOC. There are TOC tools that support the processes of strategic analysis, planning, and implementation. These include the TOC principles, the logic-tree thinking processes, and the schedule-based decision-support functions of the TVA financial management system. There are also specific TOC generic solutions ("best practices"), strategic combinations that can be adapted and combined to serve specific situations. These are beyond the scope of this volume, but are discussed briefly in the "Step 4" section of Chapter 6.

### Process Improvement

These projects should be scoped and implemented using the TOC logic trees, in order to get the maximum benefit from the least effort and expense.

### Range of Applicability

The TOC logic-tree processes can be used for a wide range of projects, including establishing new manufacturing facilities, new distribution centers, new business policies, new business processes (including the distribution, factory control, and factory scheduling aspects of the manufacturer's order management process), new measurement and reward systems, improvements in management skills, resolutions of conflicts, or changes in roles within a decentralized operations or other delegated management relationship.

### Critical Chain

This project-scheduling technique improves upon traditional PERT/CPM methods by keeping project load and priorities in line with the company's goal and available resources.

## Keeping Projects on Track

Mistakes in a large project's basic assumptions can be very expensive. It is far better to identify any needed adjustments at the very beginning of a project — through thinking alone — than to use the subsequent investment of thousands or millions of dollars, as a trial-and-error process for coming to the same conclusions. In this arena, even small improvements in the efficiency or quality of thinking, consensus, and communication processes are extremely valuable.

## It's Awful When a Project Starts Wrong or Dies Too Late

Here is an example of a project that died a merciful death that came only after too much money and time had been invested in a wrong solution. The time was 1982, a decade prior to the invention and publication of the current version of the TOC logic trees.

A large telecommunications company had spent millions of dollars, including thousands of expensive management and engineering hours, in designing, documenting, field-testing, and beginning to implement a nationwide computer system. The system's purpose was to detect and allow troubleshooting of digital transmission "bit" errors on the company's high-speed digital fiberoptic and underground cable transmission lines.

There were project managers and representatives in all the major departments, with a matrix overlay of support from all the major geographic areas. There were specification development engineers, programming engineers, staff managers for developing methods and procedures, staff managers for developing new organizational structures, and both engineers and staff managers for support of the implementation efforts. The documentation filled several feet of shelving, with quite a few people on distribution and regular maintenance updates. The paragraphs in all the books had lovely numbers such as 56.5.678-271. It was a very imposing and impressive array of resources, all built over the period of a few years on the foundation of three small assumptions. Certainly, no one would question whether such a juggernaut made sense!

A manager new to the company — a freshly-minted MBA — was assigned to overcome the resistance of two "stubborn" individuals in the one "uncooperative" department that "wasn't playing ball" with the project plan. The new manager worked very hard to persuade the two "short-sighted naysayers". He talked to them on the phone, listened to them, thought carefully about what they said, wrote them letters, called meetings, and dutifully jawboned them. All the while, he learned about the company, its objectives, the various departments, their objectives, the objectives of the big project, the progress to date, and the content of paragraphs such as 56.5.678-271. The three small assumptions were implicit, buried beneath the mountain of documentation and activity.

The new manager had done time understanding and memorizing the reactor plant operations manuals of U.S. Navy shipboard nuclear propulsion reactor installations.[68] These manuals had also occupied several feet of shelving. Consequently, he was not intimidated by the sheer volume of project information, nor by its physical scale.

The new manager had also been trained within a zero-defects culture designed to prevent a first nuclear accident. This discipline included meeting all new policies or procedures by initiating a series of new and parallel thought processes. Those thinking process kept operating, as time passed and as observations and experience were gained, until he had demonstrated to himself why and over what range of circumstances the policies or procedures made sense. A rule in the Navy reactor culture was to make sure you know what you are doing and why.

Another rule in the Navy reactor culture was: A good idea is good, no matter where it came from; a bad idea is bad, no matter where it came from.[69] Consequently, he was also not intimidated by the rank or towering international reputations of all the Ph.D.s and other experts involved in the project.

You can see where this is going. One — only one — of the three small assumptions that gave rise to the project was wrong, but that was enough to make the "naysayers" right and to make the massive project a dead duck.[70] It was going to be dead anyway, eventually, because it wasn't going to do what it was intended to do. The good news was it died *before* millions of additional dollars were spent on it. The other good news was that there was a simpler and much less expensive way to get much better results. The company took that approach instead.

## TOC Logic Trees

As demonstrated in the telecommunications project, wrong projects can be killed without TOC logic trees. However, use of the TOC logic trees moves the activity from an art performed by "the brilliant few" (who may or may not know how they do what they do) to a science or, if you like, a procedure that all the rest of us can use. The TOC logic trees assist in getting the best and most honest thinking in the assessment of the initial situation (TOC current tree) and in the formulation of the plan to improve the situation (TOC future tree). The TOC prerequisites and transition trees show the connections to the current and future trees and make the planning more efficient and clear.

If TOC logic trees had been available at the time of the big telecom project, it would have been killed much more quickly and cleanly. As it was, it took quite a few meetings and far too many white papers to get it done. A single TOC future tree would have done the job, and quickly, especially if the management culture had also been shaped to care more about the business logic of capital expenditures than about the politics surrounding them.

## Airlines Computer System

One of presentations at the 1996 APICS Constraints Management Symposium concerned the use of a current- and future-reality tree prepared on a large sheet of butcher paper covering a wall. Owen Kingman, the TOC logic-trees expert who prepared the tree diagram, used different colors of marker pens for the facts about each different department in the airlines company. Owen, an internal corporate-level TOC consultant, had been asked to facilitate a meeting in which the sponsors of a computer system wanted to get approval for several additional tens of millions of dollars to match the money they had already spent. There were "naysayers" from the other departments who doubted the huge system would work. You guessed it. Before Owen, the TOC consultant, had completed all the cause-and-effect links on his TOC logic trees, the advocates of the computer system, who were watching the trees take shape, stopped the meeting and declared that the computer project had to be scrapped. In subsequent meetings, logic trees were used to develop an alternative solution that all agreed would do the job.[71]

## The Most Fundamental TOC Projects: Making Strategic Moves

A company's most important projects are those that create its strategic positioning.

## Some Projects Are More Important Than Others

If you knew that you could have just one category of projects in your company (or life) be done right, would you not want it to be the strategic planning projects? After all, if your strategy projects over the years have placed you in the position of commanding nice cash flows, you are in the position to do just about anything else you need or want to do. Absent those cash flows, and organizational life (or life) can get a little miserable. (Any wonder why TOC organizes its efforts around the money flow called TOC throughput value added, or TVA? TVA is a lot like cash flow.)

## TOC Strategy Process Tools Help with Any and All

There are a great many competing concepts and experts in the area of strategic analysis, planning, and implementation. There's Porter, Yip, Ries and Trout, Marrus, Henderson, Schwarz, Drucker, Montgomery, Burrus, Toffler, Naisbitt, and many more. Regardless of which strategic planning concept or process is selected, the principles, structured thinking processes (logic trees), and decision-support systems of the TOC management system are helpful to managements and corporate boards of directors for focusing and accelerating business planning and improvement processes.

## Understanding What You Are Seeing in Industry and Competition

As will be illustrated in Chapter 6, TOC logic trees provide a systematic means for developing a cumulative understanding of what your company is seeing in its industry and in the behavior of its competitors.

The standard imperative to understand why things make sense matches what I once learned from an experienced executive: To avoid a lot of unnecessary mistakes, assume your competitors or collaborators are smart. Do this especially when you are certain they have blundered. Keep at this until you have explained to yourself why they are smart or until you can't stand it any longer. Then do it some more. Either way, once you have figured out *why* they are smart or once you are really convinced they have blundered, always remember that they *still* may be seeing something important that you are not yet seeing.

## Scenarios: The Art of the Long View

The TOC logic trees facilitate the orderly use of scenarios for strategic planning, as in the famous Royal Dutch Shell case. This is consistent with Peter Schwarz's vision of scenario planning as described in *The Art of the Long View*.

## Avoiding Layoffs

An important part of the TOC approach to strategy is to avoid layoffs. This is because, when the challenge of managing organizations is viewed correctly, there is no conflict between a stable employment environment and profitability.

## Other Considerations

- Implementing a continuous improvement tool will not provide a competitive advantage that cannot be copied easily. Competitors can simply begin to use the tool. The management system selected by a company must lend itself readily to formulating and implementing strategically significant combinations of technologies, resources, and activities that cannot be copied easily. The TOC management system, including the TOC thinking processes, provides companies with the means to formulate and execute these combinations readily.
- The TOC throughput value added (TVA) financial management system — when combined with drum-buffer-rope scheduling and dynamic buffering for capacity testing and with quantified protective inventory and protective capacity inputs from buffer management — becomes schedule-based decision support, one half of the formula for making sound decisions.
- The TOC logic-tree thinking processes are the other half and are used with the schedule-based, decision-support, TOC computing architecture.
- The rigor of the TOC thinking processes for strategic planning combined with the right solution for establishing financial impacts creates better solutions, better hedged solutions, at greater levels of certainty, with greater consensus in shorter periods of time.
- The TOC computing systems architecture, in combination with the TOC thinking processes, also forms an excellent decision-support mechanism for use in MRP II and ERP business processes, such as sales and operations planning.
- Use of TOC thinking processes helps a company increase its effectiveness as a learning organization by providing a clear, cause-and-effect, and graphical documentary record of situations encountered, solutions proposed, and plans implemented. The hypertext Internet and

intranet World Wide Web (WWW) and other graphical technologies lend themselves to this format very well and will lead to greatly increased working efficiency and effectiveness within and across organizations.

- A company adopting TOC is able to move with sufficient speed, flexibility, and effectiveness — in enough of the areas of importance in its selected industry — to anticipate and satisfy its customers' needs.

- When the TOC management system is implemented correctly and used well, an organization identifies and serves its targeted customers and accomplishes its own goal.

- The principles, structured thinking processes, and decision-support systems of the TOC management system are a sound foundation on which managements and boards of directors can build effective corporate strategy processes.

## Process Improvement Projects

There are three ways to approach process improvement projects. Two of them are wrong.

## The Search for Low-Hanging Fruit

Wrong method number one is the "search for low-hanging fruit". There is nothing wrong with fruit, nothing wrong with low-hanging fruit, and nothing wrong with finding such fruit and enjoying it. It is a problem, though, when large amounts of wasted time, money, and employee spirit are caught up in the politics of making a big deal out of it. Here is the way it works. Someone announces a need for a process to be changed. A team is commissioned to conduct interviews and map out all the processes. A lot of time is taken with this. Eventually, when a big report has been issued and meetings have been presented, a list of suggestions is made. The portion of the suggestion list that gets implemented is the part that anyone in the operation could have told them without all the highly visible activity. The scarce time an operation had to handle improvement correctly was squandered and gave only trivial results. Thinking too small — in order to avoid *real* thinking — does not work (i.e., does not make best use of scarce time and other resources).

## Pick a Core Process

Wrong method number two is to pick one or more — maybe all — core processes and improve on them. This approach improves a lot of things that really could have waited, does not focus on the high-leverage effects, and maybe puts the firm out of business due to over-emphasis on cost reduction (instead of growth). Thinking too big — in order to avoid real thinking — does not work, either.

## The Right Way

The correct approach is to allow TOC's five-step focusing process (entity 15 on Figure 1.1), supported by TOC logic trees (entity 7a on Figure 1.1), to identify the right things for the company to be working on. In the course of this work, specific portions of several specific processes will be identified as requiring changing and as being valuable without serious side effects on the current- and future-reality logic-tree diagrams.

These changes will not just be process improvements but will be pieces of demonstrably larger business improvements — specifically, improvements in the organization's current or future TOC throughput value added (TVA) money flows. Current or future TVA always goes up in response to a process improvement project done in a company that has adopted the TVA world scale-of-importance principle (entity 17a) of the TOC management system (Figure 1.1). *Real* thinking really works!

## Critical Chain vs. Critical Path

The standard concepts in project scheduling and management are CPM and PERT. The concept of critical chain removes the implicit assumption of infinite capacity from the project management domain, just as drum-buffer-rope removed it from the factory management domain. Rob Newbold's book in the APICS/St. Lucie Press TOC Series, *Project Management in the Fast Lane: Applying the Theory of Constraints,* is currently the best reference on this subject. Software support for critical chain project scheduling is now beginning to become available.

# TVA Financial Management System

## The Need To Create and Measure Value

An enterprise must create value for its stakeholders. Many procedures have been proposed and used to guide and measure the creation of value. Examples include allocation-based absorption cost accounting (traditional management accounting), activity-based costing (ABC), activity-based management (ABM), and the so-called "economic value added" system.

All of these procedures have been useful. In other words, they each delivered certain advantages over predecessor technologies and each continues to deliver at least a subset of what an appropriate financial system should provide. However, none of these still-popular systems has provided all of the features that would constitute the correct solution for industrial decision support and financial control. By "correct solution", I am referring to an optimum combination of concepts and processes that delivers the maximum amount of the most necessary benefits, with the least unintended negative consequences and with the least consumption of scarce resources.

## TOC Financial Foundations

The TVA financial management system is formed from the following concepts and processes: The enterprise is viewed as a system, with owners who determine its goal and stakeholders who influence establishment of necessary conditions for continued system operation. For most operational purposes, the goal of the company is established as making money, or more money, now and in the future. In the context of decisions concerning potential improvements, the goal of the company is established as making more money now and in the future. The word "money" is understood primarily to mean the economic value added by the operations of the firm.

## TOC Throughput Value Added (TVA)

Of the many competing definitions of "value", the TOC throughput value added (TVA) is defined as "all the money the system generates through sales." This means that TVA is calculated as sales, less any sales discounts, less any expenditures that vary with units, any totally variable costs (TVC).

$$TVA_{unit} = sales_{unit} - TVC_{unit}$$

$$TVA_{total \text{ for an accounting period}} = TVA_{per \text{ unit}} \times \# \text{ of units sold}$$

The primary financial measurements and ratios for the company can be determined directly from TVA:

$$TVA_{total} - \text{operating expense (OE)} = \text{net profit (NP)}$$

$$TVA - OE = NP$$

$$NP/\text{investment } (I_{materials} + I_{other}) = \text{return-on-investment (ROI)}$$

$$NP/I = ROI$$

In Chapter 6, management accounting expert Detective Columbo will provide a tabular view of a sample business planning and control scenario using this measurement format (Figure 6.1).

Accountants will recognize that the TVA portion of the TOC financial system is similar to direct margin analysis, with two primary exceptions. First, direct labor is treated as period (vs. variable) cost. Second, TOC companies treat this as gospel and actually do it vs. treating variable costing as an academic exercise learned in school.[72]

The Theory of Constraints' throughput value added was introduced as "throughput", as in the "throughput of a system", by Dr. Eliyahu M. Goldratt, the inventor of TOC, as early as 1983.[73] The alternative terms, TOC value added (TVA), throughput value added (TVA), and true economic value added (TVA) are introduced here and also in my article, "TOC Management System for Manufacturers", in the April 1997 issue of APICS magazine, *APICS — The Performance Advantage.*

Dr. Goldratt made the essential contribution, during or prior to 1983, by demonstrating the importance of organizing management financial systems around the correct definition of economic value added. These new TVA terms simply place new labels on Dr. Goldratt's initial formulation in order to improve the efficiency of communications. The new terms bring the reader's thinking more quickly into the financial arena. They remove the need to clarify repeatedly for many audiences that the term "throughput" is a financial and management accounting term and not a reference to the units of physical throughput of some machine, department, or company. Through-put, financial throughput, TOC value added, throughput value added, true

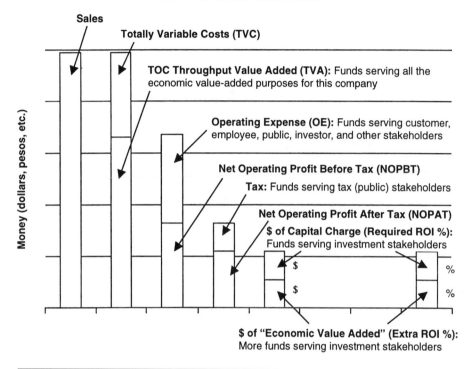

Figure 1.25. TOC Throughput Value Added (TVA), the Correct Expression of Economic Value Added, the Correct Basis for Financial Management Systems

economic value added, and TVA are all different names for the same underlying flow of money.

TVA is the economic value available to satisfy legitimate demands of *all* of the company's stakeholders, not solely the shareholders. By contrast, the primary shortcoming of the EVA system is that it measures and emphasizes the wrong portion of value that is added, overemphasizing the claims of shareholders at the expense of effects that can be observed among customers, employees, and other stakeholders in the firm (see Figure 1.25).

## Making the Pie Bigger

A management system based around TVA makes the economic "pie" bigger, without interfering with — and, in fact, supplying the funds for — any reengineering or other programs to reduce relative or absolute asset or expense expenditure levels, as appropriate during the life cycles of a company, product, or process. With the economic pie growing, these reductions in expenditure levels can be made while maintaining high levels of employment stability, without cutting corners on legal or other established obligations, and while gaining the economic, moral, and societal respect advantages of satisfying the legitimate needs of *all* of the company's established stakeholders.

The economic pie does not grow by itself. People must focus on it, plan it, do it, and succeed in their efforts. To fail to focus every day on growing the pie is eventually to back the management team into the self-defeating corner of cost-cutting without growth and to force them (in order to appear as successes in their careers) to try to persuade others that such cost-cutting is the hallmark of superior management. It isn't. Only cost-cutting within the context of growing the pie, and while delivering a reasonable degree of employment and operations stability, is superior management. Only by focusing a disproportionate amount of management attention on — call it what you will — the economic value entity pointed to by the concept of TVA will get the right thing done.

TVA focus leads to effective balance in combined marketing, product development, and expenditure control programs. By building your management system around TVA and dealing logically with operating expense and investment reduction as priorities two and three, superior employment stability, loyalty, legitimacy, respect, return-on-investment, and economic value added will come along in the process.

## Scale-of-Importance Issues

The TOC scale-of-importance principle states that not all measurements are to be treated with equal priority. The throughput world or TVA world financial policies — I'll call them policies of the TVA financial system — state that protecting and increasing TVA should always be treated with higher priority than reducing investment or reducing operating expense (Figure 1.26). Due to the higher importance of increasing TVA, decreasing operating expense (OE) should have less priority than decreasing inventory (I). Cause-and-effect thinking processes and diagrams should be used to ensure that investment and operating expense levels are understood at appropriate levels of detail. Investment and operating expense levels should be controlled and, where appropriate, reduced whenever such reduction activities do not interfere with

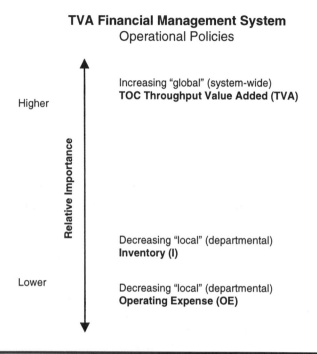

**Figure 1.26. The Ingredients of Return-on-Investment (ROI) in Order of Priority in Accordance with the TVA Throughput World and the Global-vs.-Local Scale-of-Importance Operational Policies**

efforts to increase current or future TVA. Reductions in the relative levels of operating expense and investment, as expressed by the ratios of operating expense to TVA (OE/TVA) and investment to TVA (I/TVA), should be aggressively pursued as additional desirable effects automatically coming from the programs to increase TVA.

Global (company-wide) measures take priority over local (departmental or product) measures. Therefore, once the global, company-wide TVA and operating expense are known for a specific manufacturing business plan or decision scenario, the effort to calculate allocation-based product costs is a waste of time that creates confusion and wrong decisions.

To be aware of cause-and-effect relationships affecting how consumption rates of staff (and other resources represented in period operating expense) vary with changing assumptions in the business plan is a good and important thing; to translate that awareness into allocations of overhead, to form a fictional math entity called "allocation-based product cost" is a bad, wasteful, and dangerous idea.

A company can know its costs in excruciating detail and know exactly where it stands and where it is going economically — without ever allocating a bit of overhead to form a single allocation-based product cost. Think about that.[74] Where factory department efficiency and utilization measures conflict with maintaining and increasing company-wide TVA, the global TVA must take priority.

## Better and Easier Than Allocation-Based Approaches

The TOC solution removes cost allocations from product costing and inventory valuation when making decisions. Because those calculations are not necessary for decisions, it greatly simplifies the allocations made for external financial and tax accounting alone (during the period before the external reporting requirements have not yet changed as well). In Chapter 8 of *The Goal,* the plant manager, Alex Rogo, discusses the allocations for purposes of inventory valuation with his mentor, Jonah: "But the value added to the product by direct labor has to be part of inventory, doesn't it?" The answer: "It might be, but it doesn't have to be. ... I decided to define it this way because I believe it's better not to take the value added into account. It eliminates the confusion over whether a dollar spent is an investment or an expense."[75]

Those are a few of the highlights of the TVA financial management system. In contrast to the competing activity-based and other allocation-based cost accounting systems, the TOC financial system (once seen) is clear and persuasive to industrial employees ranging from factory machine operators to chief executive officers. It is easier to use, less prone to mistakes, and less costly to administer.

## TVA's Advantages Over EVA

Compared to the economic value-added system, the TVA system's advantages include:

- Superior definition of economic value added (EVA is a subset of net profit, which is a subset of profit before taxes, which is a subset of direct margin, which is a subset of that expression of true economic value added, or TVA)
- More effective framework for accomplishing growth and maintenance of the company's revenues and cash flows

- Greater ability for inculcating loyalty, legitimacy, and productivity in the employer/employee relationship via more logical workplaces and greater employment stability
- Greater ability to fund and balance activities that meet appropriate requirements of all established stakeholders
- Greater comprehensiveness
- Avoidance of unnecessary and counterproductive conceptual and process complexity

## Inviting Detective Columbo To Help Build Consensus?

Introducing the TVA financial management system can be easy when people see that it is simpler, clearer, less prone to awful errors, and less expensive to maintain than allocation-based decision systems. On the other hand, it can be tough if people just cannot seem to shake the view of the traditional system.

It's like the "hidden pictures" game in the *Highlights* magazine found in all those dentists' offices across the country. Until you *see* the turtle that is upside-down in the lower portion of the trunk of the tree on the lefthand side of the page, you just don't *see* it. Once you *do* see it, it then becomes difficult *not* to see it. There is another well-known example in management-training circles: the picture that is either a pretty young girl or a hag, depending on how you look at it.

To help colleagues get over this hurdle, it may be necessary to arrange for a visit from that internationally recognized management accounting expert and raconteur, Detective Columbo. There is a very good chance he will pay us a visit in Chapter 6 to discuss things such as:

- TOC throughput value added (TVA)
- TVA throughput world
- TVA per constraint unit (TVA/CU)
- Life-cycle product planning with TVA-I-OE and ROI

See you there!

## Need To Know How Much You Can Make

Throughout these discussion of TVA, we have assumed we know how many units of product or service can be made and sold within an accounting period.

When making a business decision, you need to know how the capacity you have, or plan to have, will affect your decision. Conversely, you will need to know how your decision will affect the commitments you have already made against that capacity. For both reasons, scheduling technologies that deal with the finite capacities of an organization's resources are needed.

Ideally, the same scheduling technology that supports the business and strategic decision process can be used for decisions as to which improvements to select and for day-to-day production scheduling. Furthermore, ideally again, it would be best if that same single scheduling technology that serves all these interrelated purposes would also provide some previously unavailable data concerning required investments in protective inventories and protective capacities for the higher-level strategic business-decision processes.

As it turns out, and as has been described in Dr. Goldratt's 1990 *The Haystack Syndrome: Sifting Information from the Data Ocean,*[76] this is not only an ideal situation, it is also the elegant right answer. As discussed in Chapter 5, this answer is rapidly becoming the reality of a global public domain standard in manufacturing, management accounting, and manufacturing systems industries. The TOC information systems architecture described in *The Haystack Syndrome* is shown as Figure 1.27.

# TOC Logistics and Factory Scheduling

## Overview of TOC Industrial Theory and Terminology

### Factory Machines with Infinite Capacity?

Let's just tell it like it happened. The management physicist, Dr. Goldratt, stepped into the factory arena and observed the simple fact that the manufacturing computing systems were preparing schedules for factory workers under the implicit, but still very real, procedural assumption of infinite capacity. In other words, the scheduling procedures employed logic that, on its first and fundamental step, assumed that an infinite number of hours were available in each of the factory's work centers and machines. The people were finding a way to make the schedules work. They were working around the formal system. The physicist said there must be another way and set out to deduce what it had to be.

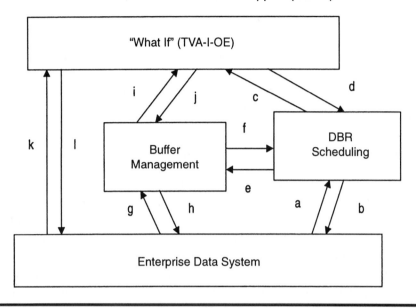

Figure 1.27. TOC Decision-Support Architecture

## Factory Machines with Finite Capacity

He began by assuming that each manufacturing company was a system. The system had machines and people who were available for some finite amount of time per day, per week, or per month. By definition, at any instant in time, given the work that the factory had to complete, only one work center or machine could be the one that most blocked the company from accomplishing its goal of making money for its stakeholders. In other words, if that primary constraint or that bottleneck were running to full capacity, it would not do any good to run other work centers faster because the primary constraint could not go any faster.

## Every Factory has a Drum, a Herbie

To illustrate this point, *The Goal* uses the analogy of a Boy Scout troop on a hike. The scout who walks most slowly, Herbie, lets all the other scouts know how fast he is walking by beating a drum they all can hear. In factories using TOC scheduling methods, the work center or machine that is the primary constraint is called the *drum*. It is also often nicknamed the "Herbie". The schedule for that department is treated as if, like Herbie's drum, it can be "heard" all over the factory (or supply chain). The schedule information for the constraint becomes the drumbeat used to synchronize all the other operations — using tight, loose, and "loose-tight" control strategies, as appropriate to each portion of the TOC logistics operation.

## Drum-Buffer-Rope (DBR) Scheduling

A factory's (or supply chain's) drum, its primary constraint, is used as the basis for its drum-buffer-rope (DBR) scheduling process. The DBR scheduling process, as one of its outputs, produces a drum schedule, the schedule of what jobs should be produced and in what order on the drum. A time element called the "buffer" is used to establish realistic commitment dates for the operation's shipments. The time element called the "rope" is used to determine the schedule for introducing additional tasks and materials into the "work in process" portion of the operation. There is a "buffer" and a "rope" time element for each task or batch in the drum schedule. Other time elements exist in the procedure; taken as a group, they are all called "time buffers".

## Synchronous Manufacturing

The DBR scheduling process synchronizes the work throughout the factory, creating a state of synchronous manufacturing by basing all other schedules on the drum schedule, the drum-derived shipping commitments, and the drum-based task and material-release schedules. It also provides feedback and control process called *buffer management*. The DBR scheduling process produces schedules for deliveries of customer orders that are reliable because they are based on finite capacity and not infinite capacity assumptions.

## Buffer Management Shopfloor Control

One work center or machine is the factory's primary constraint, its drum, its synchronous manufacturing pacing resource, its Herbie. All of the other work centers and machines in the factory are non-constraint resources or non-constraints. The buffer management shopfloor control process ensures that all these other factory resources, all the non-constraints, are working on the right jobs, at the right times in the sequence, and in the right production batch quantities in order to support the schedules on the drum and for customer deliveries.

## Early Warning System

The buffer management process monitors the upcoming shipments schedule for the shipping department, the production batch schedule for the drum work center, and the schedule at constraints-fed assembly work centers. These are two of the schedules prepared using the drum-buffer-rope scheduling process. The person or people conducting the buffer management process then note any holes in those two schedules. In other words, the buffer managers notice any upcoming shipments that might not be made on time, as well any batches of parts scheduled to be produced in the drum work center that might not be made on time — due, in both cases, to parts or other task prerequisites progressing too slowly through the operation.

## Factory Makes More Money

The buffer managers make adjustments to the schedules on the non-constraint work centers to keep from violating the schedules for shipping and for the drum work center. The reason? Because the drum schedule produced by the drum-buffer-rope scheduling process is the schedule that causes the factory to produce the most it can possibly make, consistent with the priorities in the shipment schedule (which priorities, in turn, are selected to be consistent with the goal of making money now and in the future). If meeting the drum schedule and shipment schedule established by the DBR scheduling process is the company's best and most profitable game plan — and, by definition and correct process, it is — then the non-constraint work centers do the best thing for the company and its company-level or global measurements by doing whatever it takes to support those schedules.

## Third Capacity Category: Protective Capacity

The non-constraints, both by definition and in reality, have some spare capacity. Only the primary constraint is being utilized at or near its full potential. This "spare capacity" is not "wasted capacity". It is a combination of protective capacity and excess capacity. Therefore, non-constraint work centers have two or three categories of capacity: productive, protective, and (maybe) excess capacity. If non-constraint resources are going to be able to make necessary changes and to recover effectively from any visits of "Murphy's Law" (day-to-day problems and breakdowns) to the factory, then it is not enough to have the productive capacity they need to keep up with the drum resource, which is the pacing resource, the primary constraint, the bottleneck, the Herbie. They must also have enough extra capacity, over and above what they need to support the basic production, in order to "catch up" and "get ahead" where necessary to ensure no production is lost at the bottleneck. This is because any time lost at the bottleneck (or primary constraint) is time lost from the profitability of the entire company. They must also be able to "catch up" if problems occur after bottleneck operations to protect shipping schedules.

## Time Buffers

This protective capacity works with the time buffers built into the drum-buffer-rope scheduling process to ensure no time or production is lost at the drum and that no shipments are made late. The time buffers — one each for shipping, constraint-fed assemblies, and the drum resource — are also used to calculate the times for shipping delivery dates and introductions of new materials into the factory.

## Dynamic Buffering

The process for fine-tuning the amounts of protective capacity and protective time buffers used in the factory is called dynamic buffering. This process has many other uses that are clever, interesting, and valuable but which are beyond the scope of this book.[77]

## Finite Capacity Replaces Infinite Capacity Processes

The DBR scheduling process is a finite capacity scheduling procedure. In other words, it uses facts about the finite amounts of capacity available in the company's work centers and prepares schedules on that basis. This is more realistic and more effective.

## Maximum Manufacturing Company Return-on-Investment

Taken together, the drum-buffer-rope, buffer management, and dynamic buffering factory management processes ensure that the non-constraint work centers have enough extra capacity to meet protective capacity requirements for ensuring maximum manufacturing company return-on-investment (ROI) but without having *too* much extra capacity, which would become excess capacity, which would have a dampening effect on the maximum manufacturing company ROI. They do the same to ensure that the protective time buffers — which translate to lead-time, which translates to investment in in-process materials inventories — are large enough that they do not create threats to current and future maximum manufacturing company ROI but are not so large that they unnecessarily dampen lead-times or either current or future maximum ROI.

Not too much. Not too little. It all adds up to superior ROI.

## TOC Applied to the Manufacturing, Enterprise, Distribution, and Supply Chain

The concepts of primary constraint, either material or production capacity, can be extended to include warehouses, regional distribution centers, multiple factories, and multi-company industrial supply chains. Whatever the scope of the system, the primary constraint company, factory, "white collar" department, "blue collar" work center, production machine, or material can be identified. The priorities and schedules can be formed that produce maximum system return-on-investment by making best use of the system-wide primary constraint, ensuring that adequate protective capacity and protective time buffers are in place at this higher level of aggregation. Critical chain project management scheduling techniques add a finite capacity dimension to the PERT/CPM project management methods that are popular in industry. Critical chain scheduling also offers improvements over PERT/CPM in establishing enough, but not too much, protection; this is the activity that most determines the scheduling and financial performance of enterprise and multi-enterprise projects.

## That's It

That's the overview of TOC industrial logistics theory and terminology. If you need more detail, there is plenty to be found, beginning with the resources listed in Chapters 3, 4, and 6.

### *This Has Been a Goldratt Baseline TOC (GBT)*

The TOC you get from Tom McMullen is a Goldratt Baseline TOC (GBT). That's partly because I admire Dr. Goldratt and have considered him a key mentor and friend. However, if those were the only reasons, I would not have bothered to invent the notion of GBT. I would have invented something that was more neutral and did not refer to any individual. But there are other more important reasons. The "Goldratt" part of GBT means it follows the physicist's lead in using validity, increasing validity, and superior validity as the criteria for seeking and accepting new concepts and solutions into the TOC management science knowledge base. The "Baseline" part of GBT acknowledges that no single person or organization, however much possessing of muse and genius, can cover *all* bases in all the ways that are best for *all* purposes and for *all* points of view. That means there will be innovations in TOC, in both form and substance, from many existing and new TOC experts. This is a good thing if we are prepared to deal with it properly. If we are not prepared, we will go right back to the problem we had during the bookstore tour.

Goldratt Baseline TOC was my policy as founding chairman of the APICS Constraints Management Specific Interest Group. While this may now seem the obvious policy, it was not obvious at the time. It had to be accomplished. With more and more powerful forces soon to flow, due to increasing financial and marketing stakes in the public domain TOC systems and services industries, it is still the right policy to guide TOC's expansion into a global standard management best practice.

Leaders have to lead. Sometimes, an important part of leadership is making carefully considered judgments about which portions of a large knowledge area should be brought to the attention of busy people and in what order with degrees of emphasis for further study and practice. This is especially important for people new to a knowledge area. Thinking of public domain TOC as Goldratt Baseline TOC helps with efficiently formulating, promulgating, and intelligently evolving that minimum number of simplest management concepts that started this discussion.

## Summary

### *The Theory of Constraints Management System: Emerging Global Standard*

One of the management world's best-kept secrets over the 20-year period of its grass-roots development and word-of-mouth dissemination, the TOC management system is suddenly breaking through to the mainstream of worldwide management practice.

Why? The most sweeping comment — the one that has been heard so frequently that APICS has chosen it for the theme of its annual Constraints Management Symposia — is that the TOC management system creates circumstances of culture and environment in which individuals and organizations are empowered to make common sense a common practice.

Good things happen when common sense takes over. We will have more to say about these good things in Chapter 3.

## Hits the Bull's-Eye for Managing in the Twenty-First Century

The TOC management system hits the bull's-eye for entering the twenty-first century by providing an effective combination of:

- A structured process that captures and capitalizes on combined analytical and intuitive brilliance and combined natural and trained expertise of individuals and groups
- Standard operating policies that declare that there is always a way to make progress in situations, increasing the odds of finding such ways
- Processes to identify and accomplish the high-value possibilities that exist within practical day-to-day situations
- Policies and technology that make growth in TVA a permanent high priority and thereby increase the chances of delivering stable, humane, competitive, healthy, and sensible places to spend the working part of life
- "Loose-tight" control systems that balance individual and departmental independence with organizational coherence
- A proven and growing public domain body of knowledge and experience — the TOC "best practices" — that were initiated by Dr. Goldratt, demonstrated and evolved by pioneering TOC companies and individuals, and are now supported by a wide range of commercial companies and professional societies, most notably APICS.[78]

## Where Did TOC Come From? The Short Answer

"... Most importantly, I wanted to show that we can all be outstanding scientists. The secret of being a good scientist, I believe, lies not in our brainpower. We have enough. We simply need to look at reality and think logically and precisely about what we see. The key ingredient is to have the courage to face inconsistencies between what we see and deduce and the way things are done."

         ∾ Dr. Eliyahu M. Goldratt, *The Goal*
         (Introduction, 1984 original edition)


## The Man in the Arena

"It is not the critic who counts, not the man who points out how the strong man stumbles, or where the doer of deeds could have done them better. The credit belongs to the man who is actually in the arena, whose face is marred by dust and sweat and blood; who errs, and comes short again and again (but) ... who knows the great enthusiasms, the great devotions; who spends himself in a worthy cause; who at best knows in the end the triumph of high achievement, and who at the worst, if he fails, at least fails while daring greatly, so that his place shall never be with those cold and timid souls who know neither victory or defeat."

         ∾ Theodore Roosevelt, 1910
         (Often quoted by Admiral Hyman G. Rickover, U.S. Navy)


## Virtue Its Own Reward

"Therefore, the superior man does not use rewards, and the people are stimulated to virtue."

         ∾ K'ung Chi (Tsze-sze), *Doctrine of the Mean*, XXXIII:4

# 2 Where Did TOC Come From?

The question "Where did TOC come from?" has about five answers. All five are correct, each one is different, and all are the same.[1] We will review the five answers and conclude with a discussion of the role of TOC's often-controversial originator, Dr. Eliyahu M. Goldratt.

## Answer One: Eliyahu M. Goldratt, Ph.D.

The first answer is that TOC has come originally and mostly from the muse of the brilliant and controversial physicist, educator, consultant, and philosopher — Dr. Eliyahu M. Goldratt. Here's how this view goes:

### The Early Years

There were the early years, of course, with the young Eli Goldratt[2] exercising his mind with his brother and father in memorizing and analyzing favorite classics from their cultural tradition.[3] There were more early years, which found the young Eli Goldratt now studying for a degree in physics and making waves challenging the validity of the research in all the other university departments.[4] In his spare time, he creates patents in the industrial temperature sensor and surgical instrumentation sciences.

## Breakthroughs in Scheduling

But the real story begins when Eli and his brother are asked to help a relative fix the massive production scheduling problems in a small manufacturing company, a chicken coop factory. The brothers work at it, give up, try again, make some progress, and are interrupted by Eli's sudden military service as an Israeli army sergeant on the front lines of the 1973 Yom Kippur war. War over, the brothers make more progress, begin finally to implement their computerized scheduling solution toward the end of 1974, and increase the company's production an astounding 40 to 50% by 1975 without the need to add any additional people or machines. However, the manufacturing firm fails to make necessary adjustments to policies and procedures relating to cash management and measurements. The breakthrough scheduling and production performance has allowed them to sell, make, ship, and install so many products that they have entered a situation they don't know how to handle. Their ability to *collect* cash for their shipments is outpaced by their ability to *spend* cash on materials, plunging the company into bankruptcy, whereupon Goldratt returns to finish work at the university.

## Creative Output, Inc.

By 1978, in his early 30s, Goldratt is ready to leave the university. The Control Data Corporation (CDC) likes the idea that manufacturing companies all over the world might use their time-sharing data-processing services for factory scheduling. CDC encourages the brothers to make a business out of their scheduling breakthrough of 1974–1975. The brothers proceed to build a high-flying and fast-growing U.S.-based software company, Creative Output, Inc. Some early projects are in Europe. In the first U.S. project, a General Electric defense systems unit increases throughput by 30% while halving inventory. Big contracts and rapid growth ensue. Grumman, Sikorsky, and General Motors are also among early U.S. clients. By 1980, Dr. Goldratt is making a series of presentations at APICS International Conferences. In 1983, Dr. Goldratt finds it necessary to begin to push for changes in management accounting and publishes a landmark paper entitled, "Cost Accounting: Public Enemy Number One of Productivity". That same year, Creative Output is listed by *Inc.* magazine as the sixth fastest-growing company among its "Inc. 500".

## A Business Novel, a Love Story, and a Surprise

Still, Dr. Goldratt needs a better way to help people understand the ideas behind his new management methods and software. He decides to write a book, which combines textbook, novel, and love story formats to teach his new business concepts. The book, *The Goal: A Process of Ongoing Improvement*, becomes an international best seller in over a dozen languages. A few testimonial letters arrive from people who used the ideas in the book to produce success. More letters arrive telling how the companies found bottlenecks, increased production, reduced inventories, improved lead-times and on-time delivery service to customers, increased sales, increased profits, and otherwise knocked the cover off the ball with improved manufacturing company performance. Dr. Goldratt likes the letters — until one day it hits him that companies have been succeeding who have paid only $15 for a copy of his book. The companies got the job done without his company's expensive software and consulting services. At first a concern, he soon sees this as a huge opportunity and a reminder of a larger objective established years earlier.

## Leaves One Company

The physicist in Dr. Goldratt sees this effect, assumes it will lead to a better and more comprehensive solution, and maintains the discipline of testing alternative causes. The results cause him to lose his passion for the software that forms the core of his business. He turns his attention to education, and to thinking processes. This does not please his fellow shareholders and employees in the software company. Dr. Goldratt leaves Creative Output in 1987.

## Starts Another Company

He creates another company that same year, the Avraham Y. Goldratt Institute (AGI), named after his late father. The new company, co-founded with Booz Allen alumnus, Robert Fox, mirrors the shift in Dr. Goldratt's focus from scheduling factories to the thinking of the people within organizations. The work at the new company dramatically extends the reach of his book and international reputation. He is still interested in manufacturing software, but has an entirely new class of systems in mind. More importantly, he sees things like policies, measurements, cost accounting, financial accounting, management culture, and thinking processes as far more important than the computer systems.

## Solves Cost Accounting Problem

In 1990, while working on management philosophy and policy, he also publishes *The Haystack Syndrome: Sifting Information From the Data Ocean.* Haystack describes the solution to the manufacturing industry's long-standing cost accounting problems, for which new standards in manufacturing shop control and factory scheduling are required as well. One of his most handsome and talented associates[5] demonstrates the new standard in manufacturing decision support, factory control, and production scheduling software at a midwestern U.S. division of ITT in April of 1991. "Haystack computing" begins its journey to becoming a public domain industry standard.

## Teaching the World How To Think?

Toward the end of 1991, with many important conceptual problems solved in industrial finance, production, and distribution, he shifts his focus to articulating generic thinking processes designed for use in creating solutions in marketing, sales, and engineering departments. There is also another reason he needs to invent what amounts to a new expression of the scientific method. In many situations, quagmires of implementation considerations block his existing inventions. These roadblocks stem from (1) complicated webs — different at every company and factory — of accounting and other policy constraints, and (2) the same resistance to change any new idea encounters, but more.

## TOC Logic-Tree Thinking Processes

Working with teams of people who travel around the world in the early 1990s, he develops replacements for his existing, valid, but inelegant, effect-cause-effect and evaporating-cloud analytical tools. The suite of cause-and-effect thinking tools is comprehensive, robust, and — yes — elegant. By 1994, he "test drives" this new family of logic-tree thinking processes in two high-profile applications at Bethlehem Steel and General Motors. That same year, he introduces them into the public domain via an aptly named book, *It's Not Luck.* That same year, some very smart people start telling him the new thinking processes are not only general purpose, but also "may be the most important intellectual achievement since the invention of calculus."[6]

## *Logic Trees Are New Expression of Scientific Method*

This brings Dr. Goldratt full circle from his university years. Back then, he decided to apply physics to a field outside the "hard" sciences — originally, manufacturing scheduling and management. This would be a demonstration of the idea, a means to formalize the thinking process that is physics, and a way to create a personal fortune. When it turned out that what was most needed in his chosen field of management was the formalized thinking process from physics itself, he had come full circle and nearly completed a phase of his life and career.

## *Project Management*

Cleaning out his desk as part of his first "retirement" (from active management of AGI), he is reminded of numerous "miscellaneous" innovations. One of them is a suite of Goldratt vintage late-1980s improvements to industry's popular PERT and CPM project management techniques. These form the basis for *Critical Chain,* a third business novel. Once again, the software industry takes notice of a public-domain TOC solution, and products begin to appear on the market.

## *What Is After "Retirement"?*

This brings the story to 1998. It is time for that "retirement". It is time for relocation to new quarters in Europe. For what next? Only time will tell. So, the first answer is that Dr. Goldratt did it all, which is, in a way, true.

# Answer Number Two: Intrinsic Order — TOC Was Always Right There Waiting

Answer number two is that TOC did not come from anywhere. This view suggests that, since Dr. Goldratt is a physicist, and TOC is a physics, the common-sense solutions of TOC were right there all the time, part of a pre-existing intrinsic order, patiently waiting to be discovered. This is true for TOC to the same extent it is also true for the theories of electromagnetism, chemistry, gravity, ballistics, and relativity.[7]

## Answer Number Three: Lots of People Discovered TOC, or Parts of TOC

Answer three is that, since TOC was always there waiting to be found, other people may have found parts or all of it, too. Some may have invented similar things (1) after, (2) at the same time, or (3) even before Dr. Goldratt. A variation of this view suggests that Dr. Goldratt may have found portions, or at least confirmations, of his solutions in the work of others.

The short discussion is if some concept "makes sense and works", a management physicist is bound to find it, and quickly. Since much that "made sense and worked" had already been identified and published, a physicist finds that, too, along with all the new stuff that comes from following cause-and-effect and ignoring convention. Where is the real line drawn between "new" and "already existing"? It only matters to the thinkers competing for positioning and credit. In the field, all that matters is that a management system works, and works better than the rest.

The longer discussion is if you read, for example, Peter Drucker's *Managing for Results*, one could argue that he planted the seeds for virtually everything managers and consultants have been inventing and re-inventing ever since in the areas of management science: activity-based costing (ABC), quality, the waste-reduction portions of just-in-time (JIT), continuous improvement, focusing, strategic planning, and TOC. Which raises the question, "What did Drucker read to form *his* expression of best practices in management?" Everything builds on something. Each generation of giants makes its contribution, feels not fully appreciated, and claims everything that came after was already anticipated in their own work. Each next generation of giants makes its contribution, feels not fully appreciated, and works hard to avoid acknowledging that they owe anything to the prior. I am not sure this is how it should be, but it does seem to be the prevailing practice among consultants and academics.

TOC improves upon the work of Drucker (and everyone else) in management accounting, structured thinking processes, and other areas. However, as to Drucker, (1) TOC came later, (2) TOC leaders had the opportunity to read and benefit from his work —whether they actually read or admitted they read the material, and (3) the improvements do not take away from Drucker's significant contribution to generations of managers. Same for names such as Skinner, Bishop, Hayes, Clark, Wheelwright, Levitt, Shapiro, Corey, Christensen, Peters, Ohmae, Henderson, Porter, and the like. Also, I'm not aware of industrial logistics and scheduling being something to which Drucker addressed himself in any depth. Furthermore, TOC seeks and accomplishes to be something different. TOC is fundamentally a process for inventing and re-inventing new and old processes, policies, and best practices.

Because TOC is a physics and deals only in cause-and-effect, anyone who was able to frame questions properly and persist in applying strict cause-and-effect reasoning could have invented parts of TOC, or even all of it. There might be differences in terminology. Consider Nucor Steel and Lincoln Electric, for example — companies that invented policies with much in common to TVA throughput world policies a decade before Eli Goldratt turned his attention to the matter of manufacturing management. A thinker named Rupert Sheldrake has created a concept that notes the fact that many inventions have histories showing them popping up in more than one place at once, and suggesting that it is simply because the conditions are ripe for the innovation in all those places at once.[8] Students of the psychologist Jung would suggest that messages could conceivably pass between subconscious minds sharing the same "collective unconscious". Finally, of course, if post-electronic communication is good enough for Luke Skywalker in *Star Wars*, then why not for simultaneous inventors of breakthrough ideas?

Everyone who undertakes the study and use of TOC is encouraged to invent TOC for himself or herself, to begin with the few basic TOC premises and derive all the rest of it with step-by-step logic. This means that other people may already have discovered some of the things today called TOC processes or solutions, and many more will discover and rediscover *all* of it soon.

## Answer Number Four: Many People Worked With Dr. Goldratt on Developing TOC

The fourth answer acknowledges that Dr. Goldratt is no ivory tower physicist. He stands on no ceremony, believes that physics can and should be applied — by everyone — to all aspects of life. Consequently, he knows, is known by, and has talked and thought about the various aspects of TOC with many people around the world.

It is clear that each person he speaks with is a sounding board for Dr. Goldratt. Dr. Goldratt practices what he preaches; he verbalizes his intuition by using one-on-one conversations, formal courses, and speaking engagements in order to find out — right along with everyone else — what he thinks and what is his own latest revelation on a new topic. It is not at all an exaggeration to say that Dr. Goldratt invented TOC one conversation at a time during 20 years of development.

When someone is a sounding board for someone else, when two people come into that relationship called conversation, is it only the one who is talking that does the inventing? Since he has had many employees, associates, partners, and clients over the course of over 20 years of TOC field research work, many people and many conversations have contributed to the formulation of TOC, as it is known today.

## Answer Number Five: "Graduate Students" Continue Independently

The fifth answer requires mention of a phenomenon in the TOC community I like to call the "graduate student effect". It requires acknowledgment of the many former partners, associates, employees, and clients around the world who have thought about theoretical and field research problems with Dr. Goldratt. Some believe they invented some of the TOC solutions while working with him. Some believe they have gone on to develop new and better ideas, or new and better levels of detail on older ideas. This is the way things always are in research.

As with all "graduate student" research situations, there is the first reality that the supervising professor has created the domain of activity, attracted support, and somehow funded the ongoing original research activities. There is the second reality that the "students" are sometimes re-inventing things the sponsoring professor has already invented and sometimes "inventing" things the professor has brought very close to completion and "given" to the student so he or she can succeed. There is also the third reality that the students sometimes find surprising and important new things of their own. These facts have existed within the world's many professor/student relationships across the ages. Choices are made by individual supervising professors and students concerning who gets how much of the credit for new ideas and who gets how much of any ensuing economic benefits. Differences in perception and interest can lead to differences of opinion and negotiating position. Nothing new here.

It is difficult to know whether some of the innovations made by others actually go beyond Dr. Goldratt's own thinking. In many cases, Dr. Goldratt has already been to a certain level of detail in some conversation, in some implementation site, but has not written the material down. In addition, as he describes in his 1990 book and as anyone who has worked with him can attest,

Dr. Goldratt makes a science out of provoking others to invent his solutions as their own. Leaving some details out of the literature requires others to invent the solutions, thus generating "emotion of the inventor", developing "ownership" of the solution, and increasing the chances of effective implementation. This is one aspect of what he calls the "Jonah process" or the Socratic method.[9] And, of course, all of this lofty rationale, while true enough, is also very convenient and excellent justification for rarely placing a footnote or reference into any published work. It is all a matter of style, I guess.

Many of Dr. Goldratt's "graduate students" get agitated about this issue of whether they invented something or not. To me, it doesn't matter much. I can think back to conversations or correspondence in which I used new terms or highlighted new issues, only to find them coming from Dr. Goldratt in his later presentations or writings. I have chosen to take this as a compliment.[10] I continue to appreciate Dr. Goldratt's willingness and ability to withstand both the nice benefits but also the awful demands of leadership at the cutting edge of new ideas and am happy if I can assist in some small way from time to time.

At the end of the day, what matters is the validity, promulgation, and effective application of the ideas, whether originating from Dr. Goldratt alone, from Dr. Goldratt plus some partner in dialogue, or from someone other than Dr. Goldratt.

## Bottom Line: Dr. Goldratt Has Provided the Leadership

All of these answers are correct. In fact, all of these answers are the same. No matter what else is true, Dr. Goldratt put the right frame on the questions, pointed to the right methods and standards, provided the personal example, and — there is really no question about this — did the bulk of the work of inventing and creating TOC. He provided the leadership and inspiration, did much of the marketing and promotion, and provided the attention-getting "sound bites" required for leadership communications to a busy and preoccupied industry. He either created the detailed innovations of TOC or was at least able to recognize and highlight the needed pieces of correct solutions — like any capable supervising research professor or executive — when they appeared within his partner, associate, licensee, or client base. Dr. Goldratt persisted, ground it out,[11] provoked, encouraged, contributed, persevered, accomplished, and has prevailed.

# Sharing the Process of Genius

There are several ways to think about Dr. Goldratt.

## As a Physicist

Dr. Goldratt will win some sort of prize someday, maybe a physics prize for causing demonstrations all over the world that physics is applicable outside the traditional hard sciences. Maybe it will be a prize having to do with the impact of the widespread adoption of his expression of physics methods on the ability of disparate groups to work together, in a world made smaller by revolutions in communications and transportation industries. Maybe both.

## As an Inspirational Motivator

Say what you will about Dr. Goldratt, but his will to inspire people to create change for the better — to do good, to be at their best, and to do it safely — is very clear. There are many who inspire people to action, but the focus is not always clear, nor are the safety nets. So people get all fired up, tilt at windmills awhile, lose credibility in their home environments, and end up disillusioned and less capable than before. Because of the way they tried to make a difference while inspired, they become worse off. The world, too, for the standard advice becomes, "don't try".[12] Dr. Goldratt's techniques are practical not only for the everyday things of life, but also for developing realistic courses of action in support of useful grand designs.

## As a Philosopher

In my estimation, one can consider Dr. Goldratt to be a philosopher who has provided not a replacement philosophy or world view, but rather a meta-philosophy. TOC is a set of tools that can be used to help all the other systems accomplish more of what they are already trying to do or to facilitate cooperation among the different systems, or both.

## As a Practical Kind of Guy

No ivory tower academic here. He wades right into the details in order to discern the concepts that underlie, order, and maybe, depending on your view of these things, create them.[13]

## As a Self-Directed Change Agent

Dr. Goldratt has managed many situations, but he is not really a "manager" in the sense that we think of someone in a corporate context. It is, I think, better to think of him as a self-directed change agent.

## As a Genius

I have known a lot of geniuses. Dr. Goldratt is at least as smart as any of them. Plus, he has several characteristics that make him stand out even in this group:

1. Most geniuses do not know how they do what they do. They just do it. Dr. Goldratt has made the processes of genius clear, practical, and accessible.
2. Many geniuses studiously mask the mechanics of their brilliance, in order to compete, negotiate, and dominate more effectively and completely. Having figured out what he does, Dr. Goldratt has placed his results into the public domain.
3. Some geniuses are willing to make others more capable but lack patience or teaching skill. Dr. Goldratt is a gifted educator, at both obvious and subtle levels.
4. Some geniuses honestly believe, and radiate, that the source of their genius is genetic or otherwise inherited (vs. influenced powerfully by learned attitudes, thinking processes, or other factors of environment). These people too often develop either benignly or malignantly thoughtless presumptions about the potential for genius within every other human being. Dr. Goldratt's tools bring out the best in everyone, by design.

The point is that Dr. Goldratt is a genius who has taken the trouble to understand and share the processes he uses to create breakthrough innovations. As a result, his ideas are becoming an important influence in thinking all over the world. This particular genius should be rewarded. Why? Because his input has been positive, his contribution has already run far and wide, and his story is only just now beginning.[14]

"Wishing to rectify their hearts, they first sought to be sincere in their thoughts. Wishing to be sincere in their thoughts, they first extended to the utmost their knowledge. Such extension of knowledge lay in the investigation of things. Things being investigated, knowledge became complete. Their knowledge being complete, their thoughts were sincere. Their thoughts being sincere, their hearts were then rectified."

      ∾ Confucius, *The Great Learning* (verses 4–5)

"God offers to every mind its choice between truth and repose. Take that which you please — you can never have both. Between these, as a pendulum, man oscillates. He in whom the love of repose predominates will accept the first creed, the first philosophy, the first political party he meets — most likely his father's. He gets rest, commodity, and reputation, but he shuts the door of truth.[1] He in whom the love of truth predominates will keep himself aloof from all moorings, and afloat. He will abstain from dogmatism, and recognize all the opposite negations, between which, as walls, his being is swung. He submits to the inconvenience of suspense and imperfect opinion, but he is a candidate for truth, as the other is not, and respects the highest law of his being."

      ∾ Ralph Waldo Emerson, from his essay, "Intellect"

"The errors of contemporary revolution are first of all explained by the ignorance or systematic misconception of that *limit* which seems inseparable from human nature and which [thoughtful and constructive] *rebellion* [as opposed to thoughtless and destructive revolution] reveals. Nihilist thought, because it neglects this frontier … reaches the point of justifying total destruction or unlimited conquest. … To escape this fate … it … must … draw its inspiration from the only system of thought which is faithful to its origins: thought which recognizes limits … *if the limit discovered by rebellion* transfigures everything, if every thought, every action that goes *beyond a certain point negates itself,* there is, in fact, a measure by which to judge events and men."[2]

      ∾ Albert Camus, "Thought at the Meridian", in *The Rebel*

"I like it. I like TOC. But it has nothin' to do with the kind of stuff these other three guys are talking about. I like TOC because it works."

      ∾ Mike[3]

# 3 Why Do People Say They Like TOC?

There are many answers to the question "Why do people like TOC?" because people and organizations vary so widely in their backgrounds, attitudes, wants, needs, and expectations.

## TOC Management System Makes Common Sense a Common Practice

The most sweeping comment — the one that has been heard so frequently that APICS has chosen it for the theme of its annual Constraints Management symposia — is that the TOC management system creates circumstances of culture, process, and environment in which individuals and organizations are empowered to make common sense a common practice. Good things happen when common sense takes over. For instance ...

## TOC Increases Profitability

- In April 1995, Karl Overall and Ken Papp, of the Avery-Dennison Corporation summarized the company's experience with TOC: "Bottom line, we have seen dramatic profitability and customer service improvements ... as evidenced by dramatic improvements in division, group, and corporate income statements, balance sheets, stock price, and cash flow (information is a public record via our quarterly and annual reports)."[4]

■  Gerald Danos of the Dixie Iron and Tool Company wrote an article in the March 1996 issue of the APICS magazine that described Dixie's adoption of TOC scheduling. The title of the article said it all: "Dixie Reengineers Scheduling ... and Increases Profit 300 Percent".

■  Zycon Corporation reports that "throughput ... doubled."[5]

■  The quintessential small company, Hannah's Do-Nut Shop, through the use of TOC accounting, pricing, focusing, and scheduling techniques, made year-to-year increases in sales (61% to $327,000), net profits (531% to $37,000), and return on investment (from 31% to 88%). This case study demonstrates that useful application of TOC is not restricted to very large companies.[6]

■  According to Samsonite Europe, "The improved operations created capacity that eventually was used to add another product line and to reduce the missed sales problem. The net effect was that, within one year, gross margins increased by 100% with essentially no change in operating expenses."[7]

■  Wigand Corporation, a maker of high-quality millwork for many of the fancy hotels and buildings around the world, used TOC to improve performance from large losses to large profits. Part of the plan involved instituting gain-sharing programs based on TOC throughput value added (TVA) that increased workers' pay from 9 to 18%.[8]

# TOC Exposes Hidden Capacity

■  In February 1995, at the Massachusetts Institute of Technology, Paul P. Danesi, Jr., a member of the Committee on Research of the Institute of Management Accountants (IMA) and the financial vice president of a Texas Instruments (TI) operating unit, presided at an event organized to announce the IMA's research report concerning TOC (the 1995 Noreen, Smith, and Mackey study). The event was jointly sponsored by the IMA, the Goldratt Institute, and a division of the EG&G Corporation. Danesi's opening remarks included the fascinating statement that TOC had been used to recover over $500,000,000 of hidden capacity in the company's semiconductor operations. The company was able to set aside planned capital expenditures for equipment the industry calls a "new front end". Danesi explained that, because the "new" capacity was available for use right away (vs. after a years-long procurement and construction cycle), TI was able to complete major new product introduction cycles more quickly, which provided powerful advantages in strategic positioning related to time-to-market performance. According to Danesi, the group of TI employees who implemented the TOC portion of the work were known as the "TI Jonah network".

- According to Bryan L. McKinney, Carl F. Klein, and David R. Angst of Johnson Controls, "Compared to June, the throughput at the end of November had increased over 100%."[9]
- Again, Zycon Corporation reports that "throughput .... doubled."[10]
- In 1996, after a year of work with TOC, the team at an electronic assembly plant located somewhere in the U.S. had, due to using TOC, doubled the amount of product produced in the factory with the same resources. Who was it? It's a secret. More later.
- In April 1995, Arjun Israni, a National Semiconductor division director of quality assurance wrote that, "National has been able to release substantial 'hidden capacity' as a result of the application of TOC, which has become part of the daily thinking in National factories."[11] Israni's verbal commentary included the capacity recovered could be valued in roughly the $300 million range.
- At Harris Semiconductor, Bob Murphy reports that "output increased 28%. These results occurred with relatively flat spending and head-counter levels."[12]

# TOC Makes Growth (in TVA) the Perpetual Top Priority

This is one of the key TOC policies for maintaining reasonable degrees of operational and employment stability. It will be discussed in the section that deals with saving factories and jobs. TOC does not suggest just any growth, but growth in the money flow called throughput value added (TVA). This means that product mix and its relationship to the capacity on the company's constraints are key. In fact, many of the TOC success stories occur because ...

## TOC Improves the Profitability of Product Mix

*The Haystack Syndrome* contains the famous "P&Q" product planning and process improvement exercise that has been used in countless TOC seminars. One of its points is to point out that a factory can generate more cash (TVA) if the company identifies or selects its constraints and then organizes its management accounting, product development, marketing, sales, pricing, and process improvement efforts around that choice. Many companies have profited from this approach in several classical ways.

One of those ways, as has been discussed by Zycon Corporation, is to use the strategic constraint for order-promising and as a basis for supplying calibrated amounts of protective capacity. This allows capital investments to be focused in areas that will return high current and future TVA.[13]

Ketema A&E was able to focus process engineering changes and improvements and to evaluate the incoming requests as to the effects that taking on the business would have on the factory's constraints and, therefore, on the profitability of the entire firm. Sometimes large orders at lower prices made no sense given the impact on other business and needed investments in new capacity. TOC helped guide this process toward greater profitability and improved cash flow.[14]

There are many of these examples, but let's let one more suffice, Bethlehem Steel Corporation. David Schoenen, division controller, described how his company used TOC to increase both sales and profits: "We identified our capacity constrained resource (CCR) and went to work on focusing on how to exploit it. We had to change how we marketed and sold our products to leverage the CCR. At times, what we used to think was a good productivity item on our finishing mills actually slowed our CCR down. We were actually hurting our profitability. We educated our marketing and sales people, who worked with our customers to alter the mix of items they ordered."[15]

## TOC Focuses Manufacturing and Quality Improvements on TVA

- Bal Seal Engineering used TOC to focus its Shingo quality methods toward profitability and doubled profits within three months.[16]
- ITT's Neil Gallagher, President and General Manager of the Night Vision division and an expert advocate of Total Quality Management methods within ITT's Defense and Electronics group, pioneered many

of the ways TOC methods are used to focus TQM techniques on growth in sales and profitability. His team at the Night Vision division used TOC-guided TQM to double yields, triple throughput, and reduce backshop cycle time by half.[17]

- Dresser Industries, a leading supplier of roller cone and fixed cutter bits (used to drill well bores for the oil, gas, and mining industries), used TOC to reprioritize projects. Here is a comment from Dresser plant manager, Reza Pirasteh: "Typically, in manufacturing, there are a number of projects that are ongoing which consume a substantial amount of resources. ... Projects which were not on the critical path of throughput [TVA] or did not add any value were replaced with projects that made an improvement on the entire system." Pirasteh also reported that cycle times in production were reduced dramatically without additional capital investment. In one case, the addition of several heat-treating furnaces had been recommended with a capital investment price tag of $1,000,000. TOC thinking and focusing methods were used instead to reduce non-value-added activity. The result: lead-time was reduced by 77%, removing the impetus to expend the capital. Pirasteh continues, "In all these cases, if unlimited funds were available, a possible solution could have been capacity addition via more equipment and resources. This approach, however, would certainly have hindered the creativity to eliminate the non-value-added elements and continuously improve the process."[18]

# TOC Reduces Inventories in Supply Chain

TOC's drum-buffer-rope scheduling, distribution solution, and supply chain methods provide an orderly approach to reducing a manufacturing company's inventories to a level that is an appropriate balance of profits, return on investment, resource utilization, and customer service.

- Ketema A&E experienced a 40% reduction in work-in-process inventories within the first two months. Lead-times were reduced by 30 to 60%.[19]
- Zycon Corporation reports that their "inventory ... halved."[20]
- According to Parr Instruments, "Work in process reduced by more than 60%."[21]

- Morton International Automotive Safety Products states that, "We are still very early in our use of TOC. ... But some of the improvements ... we have had in the last 12 months [include the fact that] one raw material that was maintained at an inventory of 11,000 lb. is now at an average level of 2000 lb. WIP has dropped an average of 50% between work centers. Finished goods inventory, which 16 months ago needed another storage building, can now be consolidated into a single building."[22]
- At Harris Semiconductor, "Within the first six months, Harris Mountaintop began experiencing significant results in key performance indices from utilizing TOC techniques: Line inventories were reduced 40% and output increased 28%. These results occurred with relatively flat spending and head-count levels. ... Improvements in both operational and financial metrics continue at the Mountaintop plant."[23]
- At Samsonite Europe, "... work in process inventories were down from about six weeks of sales to two weeks of sales."[24]

# TOC Establishes Excellent Delivery Performance

- Zycon Corporation reports a reduction in "cycle time ... from 3 weeks to 3 days."[25]
- Parr Instruments reported that lead-times on their pressure reactor product line were reduced by 50%, as on-time delivery percentage increased from 48 to 95%. On their calorimeter products, lead-times were reduced by more than 50%, with an increase from 70 to 98% on-time. Service parts for the two product lines had over a 50% lead-time reduction, with on-time increasing from 65 to 97%. The company reported that throughput (profits) increased during this period of change, despite the fact that production and shipping lot sizes had been reduced substantially with a consequent increase in the number of setups performed on the production equipment.[26]
- Bob Bescher, the manufacturing vice president for the Pratt & Whitney business unit of United Technologies Corporation, commented on this effect in the foreword for Srikanth and Cavallaro's *Regaining Competitiveness:* "Today's manufacturing challenge requires placing the utmost emphasis on asset management. This requires a tool with which we can control and reduce our work-in-process inventories

while at the same time improve our delivery performance. ... The concepts and techniques presented here provide a foundation for learning which allows our business unit teams to achieve inventory reductions and an increased ability to accurately predict product shipment dates."

## TOC Produces Better Returns-on-Investment (ROI)

TOC's "best practice" inventions for manufacturers include a financial management system (the TOC throughput value added, or TVA financial management system, a.k.a. "throughput accounting") that is integrated with logistics operations systems (drum-buffer-rope scheduling, buffer-management shop control, protective capacity, and dynamic buffering). Taken together, these financial and logistics systems are called schedule-based decision support (SBDS).

Taken together, these tools increase operating profits while maintaining inventories, investments in capacity, and staffing at the lowest workable levels. Here's how. When shipments go up without additional capital investments in capacity and without additional operating expenses for factory staffing, ROI goes up. When material inventories go down at the same time, ROI goes up some more. When the lower inventories and better control give rise to shorter lead-times and excellent delivery service, more and (most likely) higher margin future sales are made available to the company. This is especially true if the company is also applying TOC's standard strategic methods to remove most of the uncertainty from continued unit and margin (TVA) growth. This means ROI goes up some more. Companies tend to like it when this happens.[27]

## TOC Saves Factories and Jobs

There are lots of these stories around the world, beginning with the real story-behind-the-story from *The Goal,* which took place at a Howmet Corporation factory in the midwest. In that case, the real character behind Jonah was Eli Goldratt, while the real Alex Rogo was a plant manager named Dennis LeFevre, who later became a partner at the Goldratt Institute. Here is a more recent example of a story that came about after, and because of, the story based on Howmet.

## TOC Saves Yet Another Factory

It is August 1995 at an electronic assembly plant somewhere in the U.S., maybe Oregon.[28] The plant makes a very wide variety of assemblies for electronic systems. The plant is meeting delivery requirements with percentage on-time shipments in the high 90s, but it is losing money.

The message from the corporate office is clear: Stem the losses quickly or the company will need to rethink whether to be in these businesses. The easy way to stem the losses might have been to lay people off. Instead, the plant's managers and workers, led by a resident representative from their divisional engineering group, taught themselves TOC principles and began to apply them.

## TOC Thinking Processes

From the book, *It's Not Luck*, they taught themselves how to use current-reality tree processes and diagrams. The result was powerful — a new and shared understanding of the situation among the managers in the plant's several departments of what was happening to them. They had now left finger-pointing and unclear communication behind and had gone to work on solutions.

## The Drums of Drum-Buffer-Rope

The three material controls managers developed processes for capacity planning and constraint identification within their 17 product lines. With constraints identified, they created schedules on the constraints that allowed substantially more product to be produced with the same equipment and people. (These constraint schedules are the "drum" schedules of TOC's drum-buffer-rope factory scheduling system.)

## Buffer

With schedules on constraints, the team was able to distinguish between those expediting requests from customers they could accommodate easily and those that would involve degrees of disruption and wasted motion to the operation. (Predicting the due dates of orders on the basis of realistic constraint schedules is the "buffer" process of drum-buffer-rope.) With this information, they were equipped to negotiate honest premium prices for expedited orders that merited increases.

## Rope

The schedules on the constraints allowed the materials control managers to release to production only those raw materials that would soon be used in production. (Using constraint schedules to control the release of materials is the "rope" process of drum-buffer-rope scheduling.)

## Synchronous Manufacturing

The schedules on the constraints, combined with the increased control over raw material releases, prevented non-constraint work centers from working on materials that would not be needed for near-term shipments. This made sure that non-constraint resources were working on the same jobs that the constraints would be working on. This also prevented having raw materials used for jobs not needed soon. This prevented unnecessary wheel-spinning for expediting of raw materials from the factory's suppliers. By using the schedules on the constraints to provide the "drum beat" for synchronized operations, far fewer heroics and expediting were required to get products shipped on-time.

## Results

A year later, in 1996, the team had doubled the amount of product produced in the factory with the same resources, become profitable, avoided layoffs, smoothed out production, become a better place to work, maintained or slightly increased their on-time delivery performance, brought work-in-process inventory to nearly zero, and become a star operation in the company. And that's just the beginning.

## But It Could Have Been Different

The executives or financial people from corporate could have taken charge and said, "Lay off people representing 10% of the salary and benefit expense in the factory." They were smart not to do that. Given the plant's condition, the result would have been that the remaining people would be even more overworked (because the wheel-spinning spiral of expediting had not yet been broken). It is likely the plant would have imploded.

## Another Way It Could Have Been Different

The division executives, under pressure from the Board of Directors and public shareholders, could have said, "Reward the people whose cost accounting 'efficiencies' are high, and punish those whose efficiencies are low." The result would have been another recipe for a downspiral, as frightened employees work on anything and everything they can get their hands on in order to avoid having any cost accounting idle time. This causes people and machines to be working on the wrong jobs in the wrong sequence, which inflates work-in-process inventories and causes unnecessary shortages in raw materials inventories. The result would be a deadly expediting and finger-pointing downspiral.

# TOC Reduces and, Ideally, Eliminates the Need for Layoffs

The downsizing phenomenon of recent years has hurt the western world in important areas — morale, loyalty, legitimacy. The process did not need to be so brutal.

Did companies need to reduce the prices on their products in order to become or remain competitive? Yes.

Did companies need to reduce their expenditures for capacity, materials, and periodic operating expenses in order to continue to generate funds for current operations and future investments? Yes.

Did companies need to simplify their production, product development, administrative, and service processes in order to reduce expenditures and increase both speed and quality of serving customers. Absolutely!

Did companies need to lay people off to get this done? No.

Can we explain this horrendous downsizing experience away as a one-time historic correction, due to technological trends, that simply could not have been avoided? No.

Do we need to live in a world that oscillates from a "growth mentality" to a "cost-cutting" mentality every several years? No. We simply must add a management standard practice — a "best practice" — that makes growth in TVA a perpetual first priority.

Growth in what — sales? Yes, but no.

Profits? Closer.

Then what? Make growth in what TOC calls "financial throughput" — throughput value added (TVA) — a perpetual first priority. Why? Because TVA is the flow of funds required for everything the organization needs in order to prosper and meet all its other objectives. Let's take a measured step in the right direction. Here is proposed language for use in the policy statements that define the relationship between a company's board of directors and its management.

### Proposed Guideline From and For Boards of Directors

"Layoffs are very strongly discouraged for managing the level of reported profits. Layoffs will only be supported for reasons that could cause the company to arrive at positions involving survival-level cash flows. Managers will be measured and rewarded on their demonstrated ability to provide high degrees of employment stability and should, therefore, make every effort to maintain or increase the cash flows (throughput value added, or TVA) which make this and the satisfaction of the needs of all other company stakeholders possible."

Ira Millstein, a leading authority on corporate governance and boards of directors, wrote in 1994, "It can be a difficult balancing act — better yet, a difficult integration act — for managers, but there can be no long-term profitability and shareholder gain without an attempt to integrate fair treatment of all those who depend on the corporation."[29]

Frederick Reichheld's *The Loyalty Effect* (1996) provides some quantitative background and quite a bit of very tight cause-and-effect discussion, demonstrating that loyalty — not only of investors and the most highly desirable customers but also to and of employees — delivers strong, stable, short- and long-term economic performance."[30]

Eli Goldratt's *It's Not Luck* (1994) makes the case at a more general level. This, by the way, is enough to decide to undertake and accomplish loyalty strategies. The point is developed in a discussion of the equivalent positions on the scale of importance of the *goal* of making money (TVA) and the *necessary conditions* related to shareholder, customer, and employee stakeholders.

Dr. Goldratt's marathon presentation in Salt Lake City in April 1990 developed the relationship between the rapidly evolving measurement and logistics innovations of TOC — especially the throughput (TVA) world scale-of-importance, properly segmented markets, and protective capacity — and the economic common sense of avoiding layoffs and otherwise providing stable and sensible employment.[31]

This is the sort of thing boards of directors need to build into their compensation and oversight activities in order to support the members of managements who want to drive themselves and their institutions in this fundamentally correct direction. Institutional investors need to ask managements about how they are using growth in TVA to deliver employment stability and sound workplaces. Members of constituencies who hire institutional investors to manage their money need to let those money managers know these things matter to them. That is for both public and large private companies.

For other employers, we should at least start by making this guideline of growth in TVA a very strongly recommended "best practice" in management. And every individual alive can move the world one step in this direction, and right this minute, by resolving simply to honor and compliment those managers, executives, and boards of directors who succeed in using growth in TVA to provide reasonably stable and sensible work environments. Specifically, every individual can honor those people at least more than he or she honors and compliments those who allow themselves to be backed into situations that require layoffs and other poorly managed transitions in people's lives:[32] "Tell me how you measure me and I'll tell you how I'll behave."[33]

Last, but not least, there is Robert K. Mueller, another leading authority on corporate governance, with the inescapable best answer as to how something like TVA growth for employment stability becomes best practice (i.e., companies work at doing it because it is the right thing to do): "However, our free enterprise, democratic philosophy, structures, and context set forth the assertion that a just society, with social, contractual, and covenantal linkages, is a realizable goal. This presumes responsible obedience to the unenforceable."[34,35]

Between the current dysfunctional extreme of glorifying cost-cutters and another dysfunctional extreme of making layoffs somehow against the law is a healthy middle ground. In that place, society's reward-and-disdain systems are focused on encouraging, rewarding, and honoring those who are able to provide healthy workplaces through growth in TVA. That can be growth in TVA in one product line, replacing declining TVA in another, in order to pretty much maintain the overall size of the company. Or it can mean growth in TVA in support of absolute growth in the organization. Either way, workers, managers, executives, boards of directors, institutional investors, individual investors, and individuals can "obey the unenforceable" by carrying out their respective roles in shaping the behavior of large and small organizations into this expectation and culture.

## Growth in Cash Flow (TVA) — A Perpetual High Priority

### TOC Makes Growth the Perpetual First Priority

The TOC management system's throughput (TVA) world policy is a clear statement that, while it is always important to do things that eliminate waste and drive down expenditures, it is also always more important to do things that cause money to flow into the system. A companion idea is that it is important not to allow efforts to eliminate waste and reduce expenditures to obstruct or destroy the effort to create inflows of money for operations.

Peter Drucker's article, "Reckoning with the Pension Fund Revolution", offers perspectives on one of the most important power shifts in history, the rise of pension funds as major shareholders of large corporations. He raises questions as to what management should be held accountable for and how to structure that accountability. One of his answers sounds a lot like TOC's TVA measurements and policies. As he writes about certain German and Japanese institutional shareowners, Drucker comments that "... they do not attempt to maximize shareholder value or the short-term interest of any of the enterprise's 'stakeholders'. Rather, *they maximize the wealth-producing capacity of the enterprise* [the italics are Drucker's]. It is this objective that integrates short-term and long-term results and that ties the operational dimensions of business performance — market standing, innovation, productivity, and people and their development — with financial needs and financial results. It is also this objective on which all constituencies depend for the satisfaction of their expectations and objectives, whether shareholders, customers, or employers."[36,37,38]

## Calling All Trade Unions: Tell Managements You Think TOC Is a Good Way To Go

Union members and officials may like a lot of this. Some may like all of it but may not be in a political position to acknowledge it. We all live within circumstances influenced by history. We all must play out our roles. Still, we don't have to stop thinking, and we do not have to allow history to force us to settle for less than what makes sense in this generation, this decade, this year, today, now.

There will be a portion of the people in all walks of life (call them pragmatic do-er idealists?) who thrive on taking reasonable and measured steps in constructive directions and who will see the logic and opportunity in TOC right away. If you are in that portion of the people in the labor movement, then why not lend a hand here with a little word-of-mouth support? Every piece of the puzzle matters.[39]

# TOC Creates Better Workplaces

## What's at Stake?

Because the jobs and work lives, attitudes, and family lives of employees are affected by their experience at work, because the attitudes and lives of employees' children are affected by the words and actions of their parents, and because the attitudes and lives of all of these people add up to be the morale of the nation — with all this at stake — why do we feel it is okay to manage factories and other organizations on a basis other then the underlying physics of what makes these entities work?

In order for an airplane to fly, the right principles need to be known and incorporated into the design of the aircraft. Plus, the pilots operating it need to understand those principles of cause-and-effect and work within them. In the aircraft business, lives and confidence in technology are at stake, so we demand a physics of airplanes.

Now, I ask you, why do we think it is so different for factories and other employer organizations? Why do we think we can run our employer organizations on just any idea that pops out of the fertile imagination of some self-marketing consultant or tenure-seeking academic? Aren't lives and confidence at stake here too? Why do we not demand a physics for the management of factories and other organizations?

Stevenson and Moldoveanu, in their brilliant article, "The Power of Predictability", observe: "People will do whatever they can to find a world in which they can predict the outcomes of their actions and the consequences of those outcomes. ... That is not to say that managers should reject the various programs that promise organizational improvement. Such programs are some of the only tools for survival in an intensely competitive and uncertain world. But managers must recognize the paradox that many of those tools are in fact destroying what holds organizations together."[40]

## TOC Makes Common Sense a Common Practice

Whether they are executives, production workers, schoolteachers, parents, or children, people prefer to know that what they are doing and what is going on around them make sense.

## Things "Make Sense" When Coherence Exists

By emphasizing that local or departmental actions be viewed, planned, implemented, and measured in the context of the larger organization or system goal, TOC first creates the expectation of, and then makes progress toward, the reality of a more coherent working or living environment.

The larger or global measurements are given emphasis over the departmental or local measures. The scale of importance among the measurements is made clear in order that stakeholders in various departmental and geographical locations can cooperate effectively. The cause-and-effect logic trees make the relationships between local actions and global outcomes visually clear.[41]

## Constructing "Common Sense" in Situations

But what if common sense is in the eye of the beholder? What if each stakeholder in a situation has a different view of what constitutes common sense in a given situation?

The TOC current-reality tree processes are used to construct a view, or a sense, of a situation that is common to all the primary stakeholders. From the basis of this new "common sense" of the situation, the team moves together to form shared views of the advantages and disadvantages of potential solutions and of plans for implementing the ideas.

According to a manager in the Mystery Company's electronics plant, the self-taught and cross-functional teams of TOC implementers used their first current-reality tree process and diagram to provide themselves with a shared understanding of their situation and thereby provided an effective basis for cooperation.

## Increased Ability To Test New Ideas

The Chief Operating Officer of the C/W Companies, Bryan Bloom, asked, "How much would it be worth to you to have the ability to test and know the outcome of your important decisions ahead of time?"[42]

## TOC Focuses the Ideas of Systems Thinking

A Total Quality consultant has said, "Systems Thinking ... attempts to quantify the unquantifiable, by creating dynamic models of the interactions. While

> Verbalizing intuition is a process of discovering things that an individual or a group already — at some level — knows to be true. It involves working out the ramifications of the new, newly understood, or newly affirmed knowledge. It motivates actions appropriate to the new knowledge.

**Figure 3.1. Verbalizing Intuition, Defined**

I feel that these simulations have high value, I found them very difficult to work with a group. People's eyes glaze over when they see a network of loops or a hundred entity system model full of equations. But everybody seems to understand 'if ... then' one step at a time. And then they are willing to buy the connections between things they previously thought unconnected."[43]

## *TOC Improves Communication and Build Teams*

Johnson Controls found that "... use of the current-reality tree (CRT) ... approach based on addressing the identified core issues and discussing the obstacles that are preventing change provided a feel that the real problems might be best addressed using a team approach. This was especially important for those who may have a mindset against change using cross-functional teams."[44]

## *TOC Allows Verbalizing Intuition*
## *To Play an Appropriate Role*

At the moment, somewhere in your workplace, family, love life, town, nation, or denomination, the following statement is true: There are things holding the group back that are needless and very damaging but have become "untouchable" or otherwise "taboo" issues. Verbalizing intuition (see Figure 3.1), with the support of the TOC logic trees (Figure 1.1, entity 7a), provides the brainstorms and the planning process to move to healthier circumstances.

This one little principle (Figure 1.1, entity 7a), systematically applied, can make all the difference between effectiveness and frustration, vigor and paralysis, longevity and decay. Important note: We are not going to get into the interesting, but, for purposes of this book, unnecessary, issue of figuring out from where or from "whom" intuition comes, whether intuition ultimately can be trusted, and the like. That is another book for another day. To think of verbalizing intuition in the practical manner described above has turned out to be a huge step forward in effectiveness, with huge additional untapped potential.

## Focusing All Those Positive Attitudes

### TOC Guides and Focuses the Idealist and Empowerment Approaches

Also in every bookstore in the land is shelf after shelf of books about self-improvement. That there are always so many of these titles tells you that many of them are always being sold and read. There are many people in our companies whose attitudes toward performance have been influenced by positive take-that-hill philosophies and methodologies ranging from Norman Vincent Peale, to Ayn Rand, Jose Silva, Tony Robbins, Werner Erhard, Richard Bach, and Napolean Hill. Add TOC's logic-tree and other processes to increase the focus and effectiveness of all the positive-thinking, New Age, and other mainstream and non-mainstream empowerment approaches.[45]

### Overlaying Focus Onto the Chaos?

Even many current gurus could benefit from some focusing, it seems. John Covington, the leader of Chesapeake Consulting's TOC practice, told a story about a meeting he attended with two leading consultants he admires, Margaret Wheatley (of "chaos theory" fame)[46,47] and Peter Block (of *Stewardship* fame[48]):

> "I recently had the opportunity to work with the council of bishops of the United Methodist Church (UMC) along with Peter Block, the author of *Stewardship*, and Meg Wheatley, the author of *Leadership* and the *New Science*. Working with Peter and Meg was not only a treat, but a bit of fate; our firm has designed our strategic planning process by integrating their work with TOC principles. ...
>
> "Meg Wheatley ... has dedicated her life to a deeper understanding of organizations as systems as described by science. ...
>
> "Peter Block is a very talented man. I have his book *Stewardship* on my desk as a reference; not for clients, but for our own organization. ...

" [Still,] neither Meg nor Peter seemed to appreciate the impact of focus. Meg said that if one set up impulses into the systems that it would start changing (the ol' "butterfly flapping its wings in China affecting the weather in Cottondale, Alabama"). That's fine, but how long do you want to wait, and what is the effect of tinkering with a non-constraint? There's an effect, but perhaps not the best use of your limited resources. Our TOC tools give focus to the tinkering."[49]

The same can be said for the thousands of useful philosophies of positive thinking, systems thinking, self-help, and idealism. It is important to clarify both what can and cannot be accomplished with positive attitude alone. When the complexity of the project at hand is beyond the scope of a single individual's positiveness, however great, then techniques for dealing with the complexity become dominant in importance. When the importance and complexity outstrip the abilities of the group at hand, efficient means for gaining input from external experts are necessary.

## TOC Makes Idealist Approaches Safe and Effective

TOC is a philosophy that can be used to evaluate and guide the use of all other philosophies. Whether you know it or not, they are all there, in your company, country, state, town, neighborhood, and family. Give the people an introduction to TOC and you will give them practical tools for doing what I wrote as the back cover praise for William Dettmer's book: "If you are a veteran Theory of Constraints (TOC) implementer who has read *The Goal* and *It's Not Luck*, then read Bill Dettmer's book to deepen your understanding, increase your skills, and accelerate the pace of your TOC projects. If you're a newcomer to TOC, fasten your seatbelt now; you'll soon be making a more fundamental difference — on a greater number of more important and complex projects — than ever before." Your friends and colleagues who have been influenced by Western or Eastern idealist thinking themes will like this. You will like it, too, because it will make them more practical.

## Rebalances Power Among Types

A perennial obstacle to common sense is personality types. Companies feel they need tough pushers just to get something done. With TOC, there is no need to settle for this cultural half a loaf. Some personality types will consider this to be bad news.

The categories of legitimate reservation (CLR) of the TOC logic-tree processes, when made policy in business deliberations, level the playing field in meetings. The issue becomes the validity of the fact entities, cause-and-effect linkages, and importance scales, not the power or verbal aggressiveness of individuals. National Semiconductor's Arjun Israni stated that he liked the fact that the TOC process created a leveling effect between the company's engineers and the shop floor personnel.[50]

TOC will change the nature and content of meetings. The "practical" world will not need to settle for the merely tough; instead, the move will be toward the tough and the wise.

## TOC Empowers People To Create Their Own Best Practices

### TOC Encourages People To Invent New Best Practices

Do you have any idea how much time people spend researching best practices, especially extensive benchmarking visits? Planning trips and visiting sites, negotiating reciprocal trips — all these things take time.

If companies would shift the emphasis from cataloguing best practices that might not fit anyway over to using TOC tools to create better best practices faster, the speed would be greatly enhanced. Focus on what you are trying to accomplish with the current-reality and future-reality logic trees, and let the negative branch processing deliver to you the underlying best practice.

This is the TOC intrinsic order principle again. The best practice that works for your company is there, waiting to be found, through the use of laser-sharp TOC logic-tree processes, and right now. Or, if you like, you can wait for several months, *and* spend a lot of money, *and* owe some other company a return visit, *and* incur other costs and other risks of compromised proprietary information during the return visit, *and* maybe end up with a "best practice" that was only good in the other company's circumstances anyway, *and* (due to the huge investment the company made in getting the information) require that the effort be considered "worthwhile", *and* (again, due to the company's large investment in getting access to the ill-fitting idea) maybe even implement it, *and* maybe even require people within the company to "like it" (even if it really doesn't really fit very well).

Hey, I'm not here to tell anybody what they must do. It's your choice. You do what you think is best.

## TOC Encourages People To Propose Solutions

It's okay to propose solutions in a TOC environment. That is what the negative branch process of the logic-tree tool set is for.

## Better Solutions

According to Thomas Van Langenhoven, Senior Engineer for the Computer Sciences Corporation,[51] "Dr. Goldratt's theories ... help ... to obtain reasoned solutions without experiencing the potentially catastrophic results that occur when details are overlooked."

## TOC Creates True Learning Organizations

The TOC logic-tree thinking processes are self-documenting. The diagrams capture the learning.

## Field Research in Industrial Settings

Dr. James Cox, of the University of Georgia, led a panel discussion with representatives from EMC Corporation, Ford Motor Company, and Delta Airlines concerning how TOC logic-tree thinking processes can be used in business field research.[52]

# In Short: It Works!

That TOC works seems enough for Tom Peters. In his comment about *The Goal*, this leading management consultant noted, "A factory may seem an unlikely setting for a novel, but the book has been wildly effective."

Brigadier General Patricia A. Hinneburg, U.S.A.F. (Retired), U.S. Air Force Logistics Command, found that "the concept worked. It was beginning to pay off in the depots. As we saw the successes, we realized the entire command could use constraints management."[53]

*Success* magazine provided its readership with information about TOC. They conducted a conference in Orlando, FL, in January 1995, during which Dr. Goldratt and two military officers from the U.S. Transportation Command made educational and case study presentations. The magazine's writers believed they saw a pattern emerging in their investigations of TOC practitioners and wrote: "*Goal* readers are doing the best work of their lives."

Melvin J. Anderson, a Ph.D. at Embry-Riddle Aeronautical University described Goldratt's Theory of Constraints thinking processes as "a fully operational paradigm for decision making in complex systems."

In other words — *it works!*

## Everybody Likes Testimonals, Right?

"I hate testimonials. People use testimonials to make decisions about what to do instead of using their brains to understand what's really happening."

∿ Eli Goldratt

# 4 Who Has Used TOC?

## By Now, Too Many To Tell of All Its Uses

I t was once possible to give a reasonably complete answer to the question, "Who has used TOC?" At a point in the 1980s, one could cite the chicken coop factory that started it all, the clients around the world of the former Creative Output company, the early pioneering clients of the newly formed Goldratt Institute, and the steadily increasing number of clients served by companies founded by the former associates, partners, or employees of Dr. Goldratt.

Those days are long gone. There are now so many individuals and organizations around the world, in all walks of life, using all or some of TOC's influence and tools that the best one can do is provide an interesting and representative sample. So, that's exactly what we will do.

## Rationale for Representative Sample

The purpose of this book is not to try to tell everything anyway, as that would take quite a bit more than my budget of pages. The shelf of TOC books, proceedings, audiotapes, and papers beside me as I type this extends over 20 linear feet, and even that is just the beginning of what is available.

The purpose of this book is to tell enough of the important things about TOC to enable you, the reader, and the important people in your life to make a thoughtful evaluation of TOC and to start using it. Chapters 1 through 3, of course, have already provided information about organizations and individuals who have used TOC. This chapter will identify several additional sources of public domain TOC concept and case study materials for further study and research. It will also provide short summaries of some of the cases from manufacturing, services, military, government, and not-for-profit industry sectors.

Resources that are likely to be most useful to you are accounts in the public domain that are "straight from the horse's mouth".[1] Hence, the written proceedings and audiotapes of the annual APICS Constraints Management symposia are world-class bargains. These events have been organized by the APICS Constraints Management SIG since 1995. The records of their proceedings are a little-known, high-quality, and low-priced source of know-how from leading and internationally recognized TOC practitioners, consultants, educators, researchers, and vendors.

In addition, other professional societies, research institutions, and universities have developed TOC activities and knowledge bases. Several of these will be mentioned here.

Finally, this chapter will contain an alphabetical list of several other public domain indications of TOC activity. Since some companies are aggressively being secretive about their use of TOC to secure a time-based window of competitive advantage, and other companies have internal political battles which serve to confuse the issue of whether TOC has been successful or even ever been used (why should TOC be different in this regard?), I couldn't tell of everything I know even if I wanted to. What I have done is to provide a further compendium of public domain data points that, in combination with the other information, will round out the "due diligence" information, hints, and pointers you have in this single volume.

## APICS Constraints Management Symposia

The proceedings and audiotapes of the annual APICS Constraints Management symposia are invaluable aids for individuals and organizations seeking know-how and perspective concerning the Theory of Constraints. Each year's program has been designed to provide both introductory information for people new to TOC and advanced topical and case study information for people experienced with TOC. Taken together, the programs have been designed to create a valuable portion of the public domain TOC portion of the APICS body of knowledge. This was done by providing both overview and knowledge-in-depth presentations on TOC concepts from a long list of well-known TOC experts from industry, consulting firms, accounting firms, and computer software and systems companies. In addition, a very wide range of case study presentations were provided from companies large and small, from around the world, and from manufacturing, services, military, government, and not-for-profit organizations. The programs in each of the first three years included Dr. Eli Goldratt as keynote speaker with dozens of other featured and technical presenters. The present author and Lisa Scheinkopf served as program co-chairs. In partnership with the APICS staff, we created the objectives for the series, for each program, and each session. We invited speakers to help accomplish the objectives and made the speaker introductions heard on the audiotapes that record the proceedings.

The audiotapes and written proceedings for the events are available from the APICS bookstore at modest prices. A company serious about TOC should acquire the entire series of audiotapes and proceedings for the APICS Constraints Management symposia. These materials should first be made available to the key people coordinating the TOC management system implementation, and then to everyone else via a reference library.

These audiotapes and proceedings provide a means for "getting to know" many of the experts in the TOC consulting and education industry, and at the same time you are learning the subject area of their presentation. This is also an excellent way to get a great deal of vicarious experience with TOC tools from the case studies and the experienced consultants prior to taking the plunge yourself.

## *1995 APICS Constraints Management Symposium*

This was the first in the annual series of APICS Constraints Management symposia. The organizations and individuals listed below are among those who presented at this event held April 26 to 28, 1995, in Phoenix, AZ. The full set of audiotapes and written proceedings is available from APICS as item #01514.

- *Dr. Eliyahu M. Goldratt* — This is a 4-hour presentation concerning the history of TOC from the earliest days at the chicken coop factory through April 1995. The set of three audiotapes (APICS item #01503) is called "What Is the Theory of Constraints?"
- *Avery-Dennison* — Ken Papp and Karl Overall of Avery-Dennison provided an impressive and in-depth analysis of the changes to

measurements and operations policies at a large group within Avery-Dennison. This single audiotape (APICS item #01511) is said to have given rise to changed policies and practices in other areas of the office products industry and is entitled "What Drives Your Business?"

- **U.S. Air Force Medical Service: General Roadman, U.S.A.F.** — This was one of the most important and impressive presentations in the conference. This was a time, in 1995, when it was difficult to shake the perception that TOC was "a scheduling thing", a software program, or "something that only applied to the supervisory levels of operations and materials management portion of manufacturing companies". In a single brilliant presentation, General Roadman and his team showed how TOC principles and logic-tree thinking processes are used (1) by executives, (2) at the level of enterprise strategic planning, (3) to reengineer major activities having nothing or little to do with anything that would normally be called manufacturing, (4) by a large number of Air Force hospitals around the world, (5) by a service organization vs. a manufacturing company, and (6) by a military operation vs. manufacturing operation. This single presentation, made possible also through the efforts of Dr. Richard Moore of the TOC Center who worked closely with General Roadman on his projects, did much to help the Constraints Management Symposium's designers and the APICS Constraints Management SIG put "the right frame" around TOC for further work within APICS and its global constituencies. The 1995 presentation of General Roadman and his staff is contained in two audiocassettes (APICS item #01513) called "Insights and Observations: The TOC and the U.S. Air Force Medical Service".

- **ITT AC Pump** — During the conference, Robert Vornlocker, one of the driving forces[2] in the world's first successful implementation the TOC information systems concepts found in *The Haystack Syndrome* (April 1991), provided an overview of the implementation issues and solutions.[3]

- *National Semiconductor* — Arjun Israni, the Quality Assurance Director at National Semiconductor's WAN Division, in Santa Clara, CA, described how National adopted TOC and regained a very large amount of hidden capacity. National used TOC's five focusing steps and drum-buffer-rope in several National plants around the world and were able to increase production without the need to invest in very expensive additional capacity. They also experienced improvements in lead-times and inventories. Arjun's presentation is contained in one audiocassette (APICS item #01509) entitled "Application of the Theory of Constraints in the Semiconductor Industry".

■ *Zycon Corporation* (multi-layer printed circuit boards) — Larry Shoemaker, the Chief Financial Officer of Zycon, presented his company's experience with TOC. The shop floor personnel really like it. Zycon took the integration of strategic and operational planning with TOC to the level of selecting a strategic constraint. The company was able to organize its capital investment, product development, marketing, and sales efforts around this strategic constraint for the benefit of consistent bottom-line results.

■ *Torrington Company* — Anthony J. Toro of the Torrington Company, a unit of Ingersoll-Rand, reviewed the experience and results of his company dating back to the OPT era and through Synchronous Manufacturing with the Spectrum Management consulting organization and shared early thoughts about applications of the TOC thinking processes. The audiotape (APICS item #01506) is called "Constraints Management at Torrington".

■ *Hannah's Do-Nut Shop* — Phil and Hannah Edelman offer one of the sweetest case studies of all! This case demonstrates that TOC is well suited to very small companies. It also contains detailed financial information which can be of use to larger companies seeking to sort out the effects of a drum-buffer-rope implementation on financial accounts (APICS audiotape #01505).

■ Additional presenters included Dr. Jim Cox (APICS item #01510), Dale Houle (APICS item #01512), M.L. Srikanth (APICS item #01508), Robert Fox (APICS item #01507), Warren Foster (paper), Richard Moore, Sanjeev Gupta (paper), William Dettmer (paper), Robert Stein (paper), Debra Smith (APICS item #01514), and more. For a complete list of audiotapes from the 1995 APICS Constraints Management Symposium, contact APICS.

## *1996 APICS Constraints Management Symposium*

The organizations and individuals listed below are among those who presented at the 1996 APICS Constraints Management Symposium held April 17 to 19, 1996, in Dearborn, MI. The full set of audiotapes and written proceedings are available as APICS item #01558. Dr. Eli Goldratt served as the keynote speaker, leading a team of over 30 featured and technical presenters.

- *Dr. Eliyahu M. Goldratt* — Presented the 4-hour keynote address, "An Overview of TOC Applications for Industry". The presentation and set of audiotapes made by Dr. Goldratt at the 1995 Constraints Management Symposium developed TOC from an historical perspective, but the presentation and tapes for 1996 are organized by the current generation of TOC solutions for industrial companies. He addresses specific problems and TOC solutions in production, distribution, accounting, project management, human resources management, marketing, sales, and more (APICS audiotape set #01535).

- *McMullen Associates* — This paper (no audiotape) was an impromptu, last-minute, diving-catch presentation, made by one of the most handsome of that event's male symposium co-chairs (one of the few people at the time who could give that talk impromptu) to fill in for a speaker who had to cancel at the last minute. The paper is entitled "The Systems Industry is Giving Customers a Choice: Traditional and TOC Systems". This refers to the choice of either well-integrated traditional "closed-loop" MRP/CRP or well-integrated *Haystack Syndrome*-style TOC for manufacturing systems. This paper makes the case for a systems co-standard in the manufacturing systems industry. It also points out that we are really dealing with *de facto* co-standards. This paper is also available in *Selected Readings in Constraints Management* (APICS item #05021).

- *Harris Semiconductor* — Bob Murphy and Sharon Sines delivered an important paper, presentation, and audiotape about the TOC work at Harris Semiconductor. This work is unique and important because it is one of the earliest cases of an executive team working with, struggling with, learning lessons of, and eventually succeeding with the TOC logic-tree thinking processes. Not only that, the logistics system work at the Harris factory at Mountaintop, PA, led by plant manager, Bob Murphy, turned in some very impressive results. The title of this important presentation is "Breakthrough Performance in the Semiconductor Industry" (APICS audiotape #01536).

- *U.S. Air Force Medical Service: General Roadman, U.S.A.F.* — This is an update to the presentation made by the remarkable General Roadman and his staff at the 1995 APICS Constraints Management Symposium (APICS audiotape #01537).

- *Johnson Controls: Corporate Staff and Factory Applications* — The corporate research department of the Johnson Controls company has been studying the applications of TOC for the past several years. Three Johnson Controls Corporation employees described three of their TOC implementation projects:

- Dr. Bryan McKinney, Director of Central Research
- Carl F. Klein, Manager of New Technology
- David R. Angst, Project Manager

The Johnson Controls Central Research Laboratories (CRL) found that using a TOC current-reality logic tree for cause-and-effect analysis of staff survey results was much more useful than tabulating histograms of frequency plots of responses: "The use of the CRT helped to establish the relationships between the responses and show how one issue may be easy to address, yet not make the rest of the problems go away. Using the approach based on addressing the identified core issues and discussing the obstacles that are preventing change, provided a feel that the real problems might be best addressed using a team approach. This was especially important for those who may have a mindset against change using cross-functional teams." The TeamLinden manufacturing operation (automotive facility) in Tennessee makes and ships over 65,000 finished components each day. The plant's overall productivity increased significantly during the early phases of the TOC implementation (20% in the first four months) and have been sustained. Due to better focus on constraints, the plant was able to ship 80% more product with the same labor force, except that the third shift of mostly temporary employees became no longer necessary to make up for shortages from the two day shifts. The Ann Arbor manufacturing operation (an automotive component factory) used current-reality tree processes and diagrams to gain consensus among plant personnel concerning the causes of shortages at final assembly and other logistics problems. They found that the management's over-emphasis on inventory reduction was one of their largest obstacles to improved performance. By distinguishing between wasteful inventory and strategic inventory that serves an important purpose (e.g., protecting the constraint), they were able to increase the production rate by over 100% within 6 months, using the same labor force, and still make significant reductions in inventory (APICS audiotape #01552).

- *Delta Airlines: Thinking Process Applications* — I recommend this conference paper and audiotape very highly. Owen Kingman, the talented corporate consultant at Delta Airlines, has presented interesting case studies in several settings. He has a tremendous amount of common sense, people skills, and experience with the TOC logic-tree thinking processes. His talk contained indications of the types of important decisions that can be made using TOC principles and logic-tree processes. He tells the case of the computer system that was supposed to solve a large problem. Much money had already been spent on the system. Much more was to be spent. When Owen was called in to help the several powerful and contending groups sort out questions of whether any additional money should be spent, he used a 6- by 12-foot current-reality tree, with a different color of marker used for each different department. The result was that the people who were championing the additional investments decided for themselves to shut the system project down. Much money was saved. The simple TOC logic diagram made it clear to the authors of the system that the system could not possibly fix the problem. The problem was in the structure of the organization for the task. The organization was changed and the money for the system was saved. Also described were applications to the productivity of a ticket refunds department and to food-service logistics, in addition to many other practical tips and techniques on using TOC thinking processes to facilitate group problem-solving (APICS audiotape #01548).

■ *United Air Lines: Synchronous Remanufacturing* — Barbara Paresi and Thomas Atkinson presented a case study involving use of TOC at a remanufacturing facility, an engine overhaul facility, in San Francisco. Synchronous remanufacturing was implemented (APICS audiotape #01556).

■ *Ketema A&E Corporation* — Ken Wilson discussed his company's implementation of a TOC scheduling system based on the book, *The Haystack Syndrome*. Ketema was able to reduce inventories by 40% over a period of a few months. They were able to use their Haystack-compatible manufacturing system for decision support in connection with focusing setup reduction and process improvement efforts, and with determining price and availability for quoting new orders (APICS item #01550).

■ *Parr Instruments* — Stacey Moon has risen through the ranks at Parr Instruments, from shipping clerk to production and inventory control manager. Her TOC work at this medium-sized maker of medical instruments resulted in dramatic increases in output, along with a 60% reduction in work-in-process, a halving of lead-times, and on-time deliveries increasing from 50% to more than 90% (APICS item #01539).

■ *Morton International Automotive Safety Products* — The Morton company, of course, is best-known for its salt business; however, within the community of those who have implemented TOC, Morton International is best known for two of its industrial subsidiaries: Morton Thiokol, the rocket engine company, and Morton Automotive Safety Products, mentioned earlier in this book. Willard A. Wayman reported results in connection with his experience as production control supervisor for Morton International Automotive Safety Products. "One raw material that was maintained at an average level of 11,000 pounds is now at an average level of 2000 pounds. WIP has dropped an average of 50% between work centers. Finished goods inventory, which 16 months ago needed another storage building, can now be consolidated into a single building. Throughput, the measure of our process improvement, is up 5 to 15% depending on the production area." (Refer to APICS audiotape #01554.) The senior management team includes two respected veteran TOC implementers, Doug Rapone and Fred Mesone, formerly high-ranking executives at Federal-Mogul Corporation. Another of Morton International's companies has been represented at other TOC conferences: Morton Thiokol's Bart Penrod presented an impressive TOC success story at a Goldratt Institute conference.

■ *U.S. Air Force Logistics Command: Lean Logistics* — Brigadier General Patricia A. Hinneburg, U.S.A.F. (retired), Air Force Captain William Lynch, and Air Force employee Joseph Black told of their experience with blending TOC with Total Quality Management and with using TOC to ensure that the reengineering and lean logistics efforts remained on target. In 1990, General Hinneburg was Deputy Chief of Staff for Logistics at the Headquarters of the U.S. Air Force Logistics Command (AFLC) at Wright-Patterson Air Force Base in Ohio. Among her major responsibilities at the time was oversight of five very large air logistics centers (ALC) located in California, Utah, Texas, Georgia, and Oklahoma. These centers were, in essence, repair and overhaul operations whose missions were to keep various Air Force aircraft and vehicles in a ready condition. Two years before, in 1988, the AFLC has embarked on a Total Quality campaign. General

Hinneburg's boss at the time, General Alfred E. Hansen, Commander of the AFLC, gave her a copy of *The Goal* and suggested that everyone in the maintenance operations have an opportunity to read it. This case study goes on to describe how TOC was used to make the Total Quality efforts more effective and how lean logistics came to be the label that the U.S. Air Force gave to their application of TOC to logistics reengineeringg (APICS audiotape #01544).

- *EMC Corporation* — EMC makes high-capacity disk drives systems for IBM computing systems. The story is told there that one of the founders, one of the senior engineers, bought a box of the book, *The Goal*, and passed copies of it around. Later, at an event managed by the AME, the influence of TOC was very clear. Still later, Avraham Mordoch joined the company as internal consultant and fine-tuned the drum-buffer-rope implementation. He made a presentation about TOC and supply chain management called "TOC and the Inherent Problem of Supply Base Management" (APICS audiotape #01546).

- Additional presenters included John Covington, Dr. Jim Cox, Robert Vollum, John Covington, Donn Novotny, Al Posnack, Mark Woeppel, John Caspari, Richard Moore, Al Posnack, Eli Schragenheim, Michael Harrison, Jim Low, Dan Laskey, Ray and Dawna Hansen, Gil Spodeck, and more. For a complete list of proceedings and audiotapes from the 1996 APICS Constraints Management Symposium, contact APICS directly.

## 1997 APICS Constraints Management Symposium

Listed below are the organizations and individuals who presented at the 1997 APICS Constraints Management Symposium held April 17 to 18, 1997, in Denver, CO. The full set of audiotapes and written proceedings may be purchased as APICS item #01585.

- *Dr. Eliyahu M. Goldratt* — This one-hour presentation concerning TOC project management includes mention of the remarkable results that Harris Semiconductor produced via use of TOC project management approaches in connection with establishing a new factory (APICS audiotape #01568).

- *Boeing Corporation* — Bill Garrison spent many years as a respected and award-winning production engineer in the rocket division of Boeing prior to a fairly recent promotion and assignment as production manager of a sizable components fabrication organization. He is also an official in a condominium association. Bill's presentation, "Confessions of a Self-Taught TOC Practitioner", covered a highly successful drum-buffer-rope implementation in the factory and several TOC logic-tree case studies both on and off the "day job".

- *Boeing, Printronics, TOC for Education, Chesapeake Consulting* — A comparison of TOC thinking processes with other popular structured approaches from Total Quality and other continuous improvement approaches was provided by Marcia Pasquarella, Bob Mitchell, Kathy Suerken, and Lisa Scheinkopf, respectively.
- *Hermann Miller* — Provided a furniture company case study, with Marc Groulx and Donn Novotny.
- *Dresser Industries* — Reza Pirasteh, Plant Manager, and Glenn Camp, Division General Manager, explained how their Dresser Industries factory improved performance using the Synchronous Manufacturing variation of TOC.
- *P.L. Porter* — Glenn Hobbs is a pioneer in TOC and TOC systems built to the standards outlined in *The Haystack Syndrome*.
- *Bal Seal Engineering* — Dave Pickels and Hugh Cole presented a case study of how this company combined TOC with Shingo manufacturing improvement methods. Bal Seal had made excellent progress in quality and waste reduction over a period of two years. When they added TOC to the management system, the company doubled profits within three months and set the stage for continued growth in sales and profits.
- *ITT Night Vision* — This is yet another Neil Gallagher story. In January 1992, in Orlando, the Institute for International Research (IIR) organized a two-day conference concerning the Theory of Constraints. I was asked to be program chair and to deliver the technical session concerning TOC manufacturing information systems. One of the speakers on the program was Neil Gallagher of ITT. At the time, Neil was in the Quality Department. His presentation concerned the excellent results ITT Cannon had achieved via a judicious blending of TOC and Total Quality. In the 1997 case study presentation, Neil told of his successes with TOC as General Manager for the ITT division that produces night vision lenses for military customers. He tells of breaking physical and market constraints in sequence by applying the TOC methods appropriate to each business condition.

- *U.S. Navy and U.S. Department of Commerce Center for Excellence for Best Manufacturing Practices (BMP)* — The Navy and Commerce Department, in partnership with the University of Maryland, have created a organization called the Center for Excellence for Best Manufacturing Practices. Bob Jenkins of the BMP team told the story of how, on one site visit, the team encountered a group that informed them of the status of one of their work centers by saying, "There's our Herbie!"

- *U.S. Air Force* — Kirsten Ulowetz reported on the TOC work performed at her military base. She told how the task teams determined which TOC principles and logic-tree processes to use for specific important tasks and discussed examples.

- *Bethlehem Steel* — David Schoenen, controller of Bethlehem Steel's Sparrows Point division, reported on the TOC implementation assisted by Dr. Goldratt. This case has a special place in TOC history since it was one of the two major field tests of the TOC thinking processes conducted by Dr. Goldratt when they were new in 1994 (the other is known to have been the Cadillac Division of General Motors).

- Additional presenters included Dr. Richard Moore of TOC Center (overview of TOC logistics), Scott Robertson of Spectrum Management (overview of TOC measurements), William Dettmer of Goal Systems International (overview of TOC logic trees), Ray and Dawna Hansen of Hansen Associates (breaking market constraints), John Darlington of Allied-Signal (success with scheduling software from Scheduling Technology and with TOC accounting in the United Kingdom), John Covington of Chesapeake Consulting (TOC and supply chains), Jeanine Wilmot of Deloitte & Touche South Africa (TOC and reengineering), Neville May of Acacia/Computer Associates (TOC scheduling), Jim Low of i2 Technologies (TOC supply chain systems), Richard Ling of Ling Associates (combining sales and operations planning with TOC), Tom McMullen of McMullen Associates (TOC and strategic planning), Soundar Padmanabhan of Thru-Put Technologies (TOC scheduling), and Kent Newtown of MEMC (TOC in the semiconductor wafer industry). For a complete list of audiotapes from the 1997 APICS Constraints Management Symposium, contact APICS directly.

# Additional Sources

## APICS International Conferences

Though the annual APICS Constraints Management symposia (held in April) are international, they are different from the annual APICS International Conference held each October. The Constraints Management symposia have 500 to 600 attendees, while the APICS international conferences have several thousand attendees. The Constraints Management symposia are about TOC only, whereas the October conferences include TOC and many other topics. Audiotapes and proceedings of the TOC portions of the APICS international conferences are available, just as for the Constraints Management symposia. Consult APICS or one of the TOC web sites mentioned in Chapter 6 for a listing.

## Institute for International Research

The Institute for International Research (IIR) is an organization based in New York that organizes and conducts programs on emerging concepts. The IIR asked me to help organize their January 23–24, 1992, TOC conference entitled "Theory of Constraints: Managing in the Thoughput World". The event was held in Orlando, FL.[4] I was asked to be the conference chairman and to be one of the 16 technical session presenters (my topic was "TOC Information Systems for Manufacturers"). At the time, in addition to my activities in consulting and education, I was attempting to correct the scarcity in public domain TOC literature by publishing a magazine on TOC.[5]

The published proceedings and audiotapes from this event are available directly from the Institute for International Research, 437 Madison Avenue, 23rd Floor, New York, NY, 10022; phone (800) 999-3123; fax (212) 661-6677. Unlike the TOC audiotapes for both the annual APICS Constraints Management symposia and the annual APICS international conferences (which are described below and can be purchased for individual sessions of interest), the IIR audiotapes for this event are available only as a complete set of the bound proceedings plus all of the audiotapes.

- *Kent Moore Cabinets* — Kent Moore, founder and president of the Dallas, TX, cabinet manufacturing company which bears his name, tells of moving from constant management expediting and firefighting to a very competitive, profitable, and smooth-running operation. Kent gives a dramatic indication of how smoothly things are running — now that his management team routinely uses the principles and thinking processes of the TOC management system — by showing the audience his left hand. It becomes clear that, while Kent's right hand is well-tanned, his left hand remains very pale. He explains that this is because he now has considerably more time to wear a golf glove on that very pale left hand. Shed not a tear for Kent Moore. He's doing okay. The Kent Moore story caught the attention of the furniture industry and was told in the October 1991 edition of the industry's periodical, *Furniture Design & Manufacturing.*[6] Thank you, Kent, for your generosity and effectiveness over the years in getting a lot of important information into the public domain about TOC and for taking the trouble to teach people what it is all about.

- *Stanley Furniture* — The contribution of TOC consultant and former Goldratt Institute associate, John Covington, includes a paper co-authored by Larry E. Webb, Jr., Executive Vice President and Chief Operating Officer of Stanley Interiors Corporation of Stanleytown, VA. The paper reads, in part, "Stanley Interiors, a nearly $200 million retail furniture manufacturer, began their successful implementation in August of 1990."

- *ITT Aerospace/Communications* — Neil A. Gallagher, in his then capacity as Vice President of Operations for the ITT Aerospace/ Communications Division, makes a presentation on how necessary TOC is to keep Total Quality Management (TQM) programs focused. The title of the talk is "Why the Theory of Constraints Is Essential to the Success of Any TQM Program". Neil is a veteran TOC implementer whose experience includes a pioneering, successful OPT-era implementation. His successful TOC implementation at the ITT Night Vision division is addressed in the section concerning the 1997 APICS Constraints Management Symposium.

- *Binney & Smith* — Joe Roberts, Director of Customer Service, and Dan Tretter, Manager of Operations Systems and Planning, are two managers from Binney & Smith, the Crayola crayon company. Their presentation relates the details of how TOC was integrated within Binney & Smith's home-grown continuous improvement system called "high-velocity manufacturing".

- *General Motors* — William D. Mavredes, the Divisional Administrator for Quality Network and Synchronous Organization for the A.C.

Rochester Division of General Motors, tells of the earliest Jonah courses (including the first one) and other Goldratt Institute educational initiatives. Impressive operational results are reported from application of the TOC at several facilities.

■ *AT&T*— Thomas R. Henry, Sr., served in nearly every major manufacturing role during his 28-year career with AT&T, including an assignment as Manufacturing Systems Engineering Manager at AT&T's printed circuit board facility in Richmond, VA. In a presentation entitled "How the Nation's Largest Printed Wiring Board Manufacturer Was 'Hot Wired' for TOC", Tom relates how the facility combined TQM, just-in-time, and other concepts to create a dramatic turnaround at the factory. The Richmond factory was awarded the Automation Forum's Renewal Award and won second place in the competition for the U.S. Senate's Productivity Award.

■ *Ford Motor Company* — Marc A. Lorrain, a division staff manager at Ford Electronics, tells how a Ford factory in Canada improved performance using TOC logistics concepts. As Manager, Continuous Flow Manufacturing, for Ford Electronics, he was introduced as having 4 years of experience overseeing the design and implementation of time-reduction programs using TOC and Continuous Flow Manufacturing principles in the Electronics Division and other components of Ford Motor Company.

■ *McMullen Associates* — TOC decision support, shop control, and scheduling systems, circa 1992, from Tom McMullen.

## Institute of Management Accountants Report

### Theory of Constraints Field Study Report

The research arm of the Institute of Management Accountants (IMA) is called the IMA Foundation for Applied Research. The Foundation co-sponsored, with Price Waterhouse (Paris), a project to investigate implementations of TOC in the U.S. and Europe. Their report was published in 1994 as *The Theory of Constraints and Its Implications for Management Accounting.* Chapter 3 of this IMA book (which is available from the APICS bookstore) is devoted to seven case studies, including:

■ *Samsonite Europe N.V.* — This case is also said to be the basis for the story of the "I Cosmetics" company in *It's Not Luck.*

■ *Baxter-Lessines* — This is a Belgian operation in which the TOC logic trees are reportedly used for meetings concerning really tough problems, especially those concerning personnel. Their experience has been that little gets accomplished in meetings of this kind without the use of the TOC trees. Baxter-Lessines had been very active in their Total Quality efforts. TOC provided guidance as to which quality improvements would really make a difference, which to do right away, where to focus scrap reduction efforts, etc. (see page 68 in the book).

■ *Western Textile Products* — This is the case of a U.S.-based textile manufacturer, for which Dan Lilienkamp has been a prime mover.

■ *Company J* — A manufacturer of tools for cutting metal. This company and the next chose not to be identified in the IMA report, but I include them here as an indication of the range of companies to which TOC can be usefully applied.

■ *Company Q* — A small European maker of capital equipment for the canning industry.

■ *Hofmans Forms Packaging* — A printing company. This company is said to be the basis for the "Pete's printing company" story in *It's Not Luck*.

■ *Kent Moore Cabinets* — Kent Moore and his management team have been very generous with their time in sharing what they have learned about TOC with the general public. This, of course, includes their competitors. As can be seen from his biography, Kent is that sort of individual: smart, tough, community-spirited, active, and possessing of a wide-ranging perspective. This case study shows several fine examples of TOC tree diagrams provided by members of Moore's management team.

## IMA Global Solutions Conference Series

A videotape of Global Solutions III, the third in the (then) National Association of Accountants' series, is, at this writing, available for rental by IMA members.[7]

## International Society for Systems Improvement

The International Society for Systems Improvement (ISSI) is an association formed by TOC academics and practitioners. From time to time, it issues a journal and conducts a conference. Many of its leaders and members are also members of the APICS Constraints Management SIG. There was a point in time (June 1994) when key leaders within the TOC community had to make a decision as to whether to devote time and energy to what was then only the idea of an APICS Theory of Constraints SIG or to the re-invigoration of an already-existing independent ISSI. Many, including the present author, decided that the APICS initiative would be the more sound strategic move for APICS and industry. The rest, as they say, is history. For further information, contact Dr. Stan Gardiner, Director, International Society for Systems Improvement, Auburn, GA.

## Clemson University

During the past several years, Clemson University's Office of Professional Development has produced a series of TOC conferences. Case studies have been presented by organizations including Ford Motor Company, Rohm & Haas, Rockwell Automotive, West Tape and Label, Weisz Graphics, Zycon Corporation, and Sumter Cabinet Company. For further information about the school's TOC activities, contact the Director of Training and Development, Office of Professional Development, Clemson University, P.O. Box 912, Clemson, SC 29633.

## Other Public Domain Indications of TOC Applications

The TOC applications around the world not mentioned in this book outnumber the ones mentioned by a wide margin, but here a few more.

- *Automotive Glass Industry* — I have spent time in a large glass factory that has implemented TOC accounting and scheduling concepts. As far as I know, this company has never reported its experience in the public domain. However, one indication that there has been long-standing interest in TOC in this industry is the July 1989 *Management Accounting* article, "Pricing Strategy in the Automotive Glass Industry". This article calculates TOC throughput contributions and discusses the strategic ramifications of using them in place of traditional allocation-based product costs.
- *APICS, The Educational Society for Resource Management* — APICS, the large global professional society has used TOC management

philosophies, thinking processes, and scheduling techniques for certain aspects of its self-management. The APICS Constraints Management SIG was formed from two simple TOC trees, a simple TOC future tree and a simple prerequisites tree. They were accompanied eventually by detailed proposals and business plans, but they served as powerful visual summaries of the vision and plan for the SIG operation. TOC future, prerequisites, and transition trees have been used by Bob Mitchell and Bill Garrison to design and implement new management processes for the Constraints Management SIG. Transition trees are used as instructor guides for APICS Theory of Constraints workshops and courses. Several inputs into discussions of the role of the SIGs in APICS strategy were base on TOC principles and analyses and were adopted. The APICS Educational and Research Foundation has used TOC trees to explore the nature of the research function. Dr. Jim Cox reviewed the TOC logic trees prepared by Tom Cook, an APICS past president.

- *Avery-Dennison* — Avery-Dennison's use of TOC is the subject of a paper and audiotape from the 1995 APICS Constraints Management Symposium. In addition, Steve Buchwald, an Avery-Dennison corporate consultant with a background in activity-based accounting, placed a very useful paper into the 1994 APICS International Conference Proceedings (APICS item #04068, pages 635–637). The paper demonstrates that using TOC throughput value added (TVA) accounting provides faster, easier, and more accurate monthly and quarterly financial closing procedures. More importantly, the operating personnel are able to use the TOC reports to make better decisions more quickly and with greater consensus among affected departments. Steve's papers on integration of TOC and activity-based methods also appear in various APICS regional overseas conference proceedings.

- *Dixie Iron and Tool* — Gerard Danos, the General Manager and Operations Vice President for Dixie Iron Works in Alice, TX, reported in the March 1996 edition of APICS magazine that "at Dixie Iron Works, we have institutionalized the Theory of Constraints concept." Part of the institutionalization work was to implement a TOC software system called Resonance from Thru-Put Technologies of San Jose, CA. "After only four months of running the software, we had more than doubled our due-date performance to 65% from 30%. We also more than doubled our inventory turns to 12 per year. Most significantly, operating profit increased fourfold over two years ago when we operated in our old mode, and 2.4 times over using constraints-based scheduling manually." A few months earlier, Mr. Danos had made testimonial presentations on behalf of Thru-Put Technologies at the 1995 APICS International Conference.

■ *EG&G* — An EG&G manufacturing division in the U.S. was saved from a generally positive, but still dangerously overzealous, just-in-time and Total Quality implementation by an application of the common-sense principles of TOC. There were paradigm shifts in terms of thinking and decision processes. A few, but very damaging, policies concerning scrap, batch-sizing, and work-center measurements were adjusted from a purist and self-defeating "zero-tolerance" form of implementation to a view that reflected the underlying economics of the situation. This was an introduction of cause-and-effect in place of simplistic and purist "best practice" thinking. This was also an introduction of decision processes centered on protecting and growing current and future throughput value added `to replace decision processes that were implicit but could be demonstrated to be based on concepts of eliminating any and all scrap, inventory, and perceived wasted time at any cost. There was a paradigm shift at the logistics and scheduling level. While activity-based costing, process management, just-in-time, and Total Quality Management concepts had been used for several years and accounted for a very great many person-hours, the new cellular and focused factories were not able to reduce lead-times, improve delivery performance, and shrink inventories as planned. Introducing semi-custom TOC drum scheduling and buffer management computer systems made dramatic improvements in the synchronization of the factory's work centers, boosted on-time delivery performance, replaced the "feast or famine" scenario in work centers throughout the factories with leveled production workloads, reduced overtime, increased measured productivity in most work centers, and allowed (due to the superior, yet simpler, TOC scheduling methods) several scheduling personnel to be assigned to new value-added duties. This was all done within six months of beginning with TOC. This case study was presented at the June 1994 International Society for Systems Improvement (ISSI) conference.[8]

■ *Elwood City Forge* — In February of 1996, Dr. Kevin Handerhan, president of Elwood City Forge (ECF), made a presentation to the

APICS Pittsburgh Chapter about his company's experience with TOC. They used a combination of TOC focusing processes, logistics concepts, measurements, and structured "tree" processes to create integrated marketing, manufacturing, and quality performance that restored them to a strong financial position. Work had already been underway for years on a Juran form of a quality-improvement program, but they decided to use a TOC business approach to help them "focus their problem-solving efforts (we want to solve the most important problems in terms of constraining our ability to make money and generate cash)." This meant that the quality tools would be focused on the most important problems, instead of being spread ineffectively over all problems. All of the quantitative quality, service, and financial measures improved dramatically. Moreover, Mr. Handerhan and Mr. Hamilton began to receive glowing letters from their customers and prospects.[9]

- *Flow International* — Flow International's Manufacturing Vice President, Michel Gadbois, made testimonial presentations on behalf of the company's scheduling software provider, Thru-Put Technologies, at the 1995 APICS International Conference. Later, in the March 1996 edition of the periodical published by the Institute of Industrial Engineers (IIE), Michel wrote that from 1993 to 1995, sales for Flow, a maker of high-pressure water cutting systems, had grown from $42 million to $110 million. He found it necessary to replace the MRP scheduling and chose to do it with Thru-Put's TOC finite-capacity scheduling system, Resonance. "We've reduced WIP by 50% and finished goods by 25%. We're able to make more informed management decisions such as what kind of new equipment to [buy] or manpower to employ. ... We are now able to determine optimal overtime schedules and control out-sourcing expenses."[10]

- *National Semiconductor* — See the comments about the 1995 APICS Constraints Management Symposium. The contribution of TOC to the turnaround of National Semiconductor is acknowledged in a book written by the consultant who worked with Gil Amelio on the program.[11]

- *P. L. Porter* — Their story was told at both the 1995 APICS International Conference and the 1997 APICS Constraints Management Symposium.

- *Tipper Tie* — Tipper Tie, a manufacturing company based in Apex, NC, makes machinery and supplies for the food-packaging industries. Dave

Richman's ACS presentation in 1996 told how the company had used TOC to guide changes to break market constraints. They have also used TOC thinking tools for business negotiation and planning.[12]

■ *Valmont Industries* — In the May 1988 edition of Management Accounting, David S. Koziol, the factory controller, described how TOC was used to improve net earnings by 40%. Valmont/ALS, a job-shop steel fabricator in Brenham, TX, was facing "mediocre inventory turns, increasing amounts of overtime, and a recession." Koziol wrote, "Meanwhile, we were learning as much as possible about the Theory of Constraints from the American Production and Control Society publications. I was embarrassed to discover these concepts had been around for a number of years but I'd never been exposed to them in any of my accounting publications. Jeff [Wood] ... developed additional queries to better assist in the management of the inventory buffers. These queries provided production control with a detailed analysis of parts missing from the desired buffer and their current location and process time remaining. By the end of the year, the improvements as a result of the drum-buffer-rope method of operation were plainly visible. ... The actual lead-time for certain product lines had been cut to half the assumed industry standard lead-time. Due-date shipment performance had improved to the mid-to-upper 90% range. Shipment levels increased to their highest levels in our company history, yet personnel were still near the low post-layoff levels. Shop overtime had been kept at reasonable levels and shipping costs as a percentage of sales had fallen. Net earnings were up 40% better than our operating plan for the period. Return on equity was up, and cash flow was 60% greater than the operating plan. These favorable results occurred despite a 9% decrease in actual volume as called for in our aggressive annual operating plan."[13]

■ *West Tape and Label* — According to Brad Stillahn, West's president and CEO, "We're doing better than we've done since 1987 in terms of profitability." The firm had been attempting for years to raise prices and lower sales commissions. When they shifted to TOC techniques for identifying and dealing with physical and policy constraints, they focused logistics operations on certain work centers, lowered prices, and raised sales commissions.[14] West Tape gave an update in televised news programs in the fall of 1997 (CBS affiliate WUSA in Washington, D.C. and ABC affiliate WKGO in San Francisco).

■ *Whirlpool Corporation* — At the 1995 APICS International Conference, David Mealy and Sekar Sundararajan presented a case study of a new factory being built from scratch to conform with TOC design principles. The paper described how manufacturing cycle times for many models of appliance were reduced to one day, while WIP turns were increased fivefold.[15]

■ *Xerox El Segundo* — A public presentation was made at an APICS chapter in California that involved implementing drum-buffer-rope scheduling using a simple spreadsheet program for systems support. The usual results: WIP and lead-times down, as on-time deliveries go up.[16] The implementer was Sanjeev Gupta. Keep that name in mind. You'll be hearing it again in the context of being a force in the manufacturing software, systems, and services industry.

## Summary

These are not, by any means, all of the TOC cases in the world. Again, with millions of TOC books sold and many education sessions conducted around the world, there are far too many cases for anyone to keep track of. This is as it should be with an emerging management standard. What we have in this book are enough of the right different types of cases that are, to one extent or another, indicated somewhere in the public domain. This allows you and your colleagues to know that TOC is real, that it has been used in many different ways, and to know that your competitors may already be using it. Bear in mind that, while each case highlights a useful aspect of TOC that has produced impressive results, many of the cases involve TOC implementation approaches that are outdated, unnecessarily risky, or are otherwise no longer valid. The recommended and up-to-date approaches for getting impressive results from introducing TOC, but without the unnecessary risks or unintended effects, are described in Chapter 6.

## The Best or the Worst of Times?

"It was the best of times, it was the worst of times, it was the age of wisdom, it was the age of foolishness, it was the epoch of belief, it was the epoch of incredulity, it was the season of Light, it was the season of Darkness, it was the spring of hope, it was the winter of despair, we had everything before us, we had nothing before us, we were all going direct to Heaven, we were all going direct the other way — in short, the period was so far like the present period, that some of its noisiest authorities insisted on its being received, for good or for evil, in the superlative degree of comparison only."

ᴄᴡ Charles Dickens, *A Tale of Two Cities*

## So Many Challenges Are New

"So many different people. So many different kinds. For better or worse, different people."

ᴄᴡ Gwen Stefani, Eric Stefani, and Tony Kanal (of No Doubt),
"Different People" on *Tragic Kingdom*

## But True Constraints Are Few

"By nature, men are nearly alike; by practice, they get to be wide apart."

ᴄᴡ Confucius, *Analects* (16:2)

## Part of the Answer ... Belongs to You

"To believe your own thought, to believe that what is true for you in your private heart is true for all men — that is genius.

"Speak your latent conviction, and it shall be the universal sense. ... Familiar as the voice of the mind is to each, the highest merit we ascribe to Moses, Plato, and Milton is that they set at naught books and traditions, and spoke not what men but what they thought. A man should learn to detect and watch that gleam of light which flashes across his mind from within, more than the lustre of the firmament of bards and sages.[1] Yet he dismisses without notice his thought, because it is his."

ᴄᴡ Ralph Waldo Emerson, *Essay on Self Reliance*

# 5 | Where Is TOC Headed?

## Any Important and Complicated Problem Will Do

Where is TOC headed? Anyplace — and every place — that contains one or more important, difficult, and complicated problems. How can one know this? Three reasons:

1. People care about the world around them.
2. People who care will do what works.
3. TOC works.

TOC will go everywhere people take it. Because TOC is a simple, powerful, and comprehensive tool set for use in addressing practical problems in all areas of life, it will proceed in a great many different directions. Some of these directions I am sure about, because people are already in motion. Others are likely due to the nature of certain existing situations and the nature of TOC's potential contributions there.

## *Bigger and Tougher Problems Can Reasonably Be Tackled*

### *"The Wisdom To Know the Difference"*

There is an old saying, or maybe it's a prayer: "God give me the courage to change what I can change, the patience to endure what I cannot change, and the wisdom to know the difference."[2] This makes a lot of sense but leaves open the crucial question of how to apportion problems and opportunities between those categories. It also leaves open the question of *how* to influence issues too large for any one person to tackle directly themselves.

## On the One Hand, There Is Nothing New

There is rarely a question among people of good will about whether ideal situations would be nice to create and maintain. For most people, the questions go more along the lines of whether there's enough time or other important resources to make progress toward the ideal, whether there is a reasonable chance of success, whether the risks are such that it will be safe even to try, whether the risks of inaction outweigh the risks of taking action, and which other things in life will need to be set aside or sacrificed.

There is also a smaller group of people — whom we call idealists, dreamers, or radicals — who go to work toward the ideal with little or no calculation. They move sometimes to success, sometime to chagrin, and almost always at what most people would call great personal cost.

That part's not new.

## On the Other Hand, Something Is New

What *is* new is that the TOC principles and thinking processes completely change the calculations concerning what is possible and by what means. One key consideration that experienced people calculate is whether people can cooperate effectively enough. The TOC thinking processes alter this calculation dramatically by focusing time on high-leverage actions and by increasing the effectiveness of communication, empowerment, and delegation.

## More Skills Mean More Is Possible

In 1982, during my first year as an AT&T employee, a freshly minted MBA from Harvard Business School, I had a discussion one evening with my manager. I have always been good at digging into situations and, in fairly short order, formulating one or more practical breakthrough solutions. We were discussing one particular idea that he thought was overly idealistic.

He was eloquently seeking to persuade me to support the "pragmatic" strategy (which was a wheel-spinning, high-visibility, multi-meeting show of activity with only the tiniest element of true progress for all the expensive time we would spend). He was a good man who understood the game that existed there at the time, who viewed himself as not in a position to change the game's rules, and who really thought he was advising me well about balancing practical vs. idealistic motivations.

I listened respectfully and carefully for a long time. After awhile, I simply said, "The faster and more skillful you become, the closer to the ideal you can get." He thought about that for a long moment, smiled, and went to work figuring out how he was going to manage the environment around me to let me get the right job done. I used a practical creative thinking process that what we would today call an implicit version of the TOC thinking processes[3] to formulate an integrated engineering, operations, information systems, organization, job design, and measurement solution that saved AT&T and its ratepayers several million dollars.

## The Wisdom To Know the Difference: Revisited

Think about the current-reality logic tree, which organizes symptoms, intermediate causes, root causes, and core problems. This helps clarify that, even though symptoms seem more manageable and practical in the short term, it is possible to gain a working consensus on the need to address deeper causes.

Consider the future-reality logic tree, which allows a group of people to see and become committed to a shared view of a better future situation. Consider the prerequisites logic tree, which helps create a shared understanding of how the large elements of an overall much-improved future situation can be broken down into manageable pieces and then assigned to the several or many individuals, departments, or organizations needed to get the job done.

If you are not already fast and effective, TOC principles and thinking processes will make you faster and more effective. If you are already fast and effective, use TOC techniques to move "up-market" into working with more fundamental, more important, and more complicated problems and opportunities.

Two people, one with and one without skills in the use of TOC logic-tree thinking processes, will mean two very different things when they speak of the "wisdom to know the difference". Think about that.

## Best of Times or Worst of Times?

Are we living in the best of times or the worst of times? The answer, of course, is — *yes!* It always has been, is, and most likely always will be both! Sure, we are beset with problems. But how must things have looked to our parents and ancestors during the Black Plague, the Napoleanic Wars, the pioneering covered wagon rides, the First World War, the Great Depression, the Second World War, all the world's revolutionary wars, the civil wars, the religious wars, and the countless other extremely serious problems? Through it all, people worked on the problems and made progress.

For example, despite all these problems, the world population has grown from several million to several billion people! During that period of danger, disaster, and misfortune, something must have also been going right for the population to grow so dramatically and for the standard of living for many to rise so much.

While many still lack proper nutrition and health care — a problem many people are working on as this is written — more people have more than in the past, and more people will have more in the future. Once again, the number of people in the world has been skyrocketing.

Where once people of certain skin colors had no choices in one of the world's well-known countries, there are now people of color sitting on the highest court, courted to become president, and found in virtually every professional and social walk of life. Though much remains to be done on this and other issues, people have persisted in making changes that matter.

The point is that despite all the bad news, people were still causing good to happen. That is our challenge now. Amidst all the bad news, we must set ourselves to the task of making progress.

Things seem worse in some respects, better in other respects. What to do? Answer: Get to work on what could be better. So, let's do it.

If you believe you have limited skills, pick a small problem, work it, finish it, and pick another one. Do it again. At some point, move to a more difficult problem. The TOC tools will serve you as you work your way up from easier to more difficult problems.

## The World Is Smaller Now

A large part of my motivation to help build TOC into the foundation of standard worldwide management practice arises from my belief that TOC delivers an important piece of the answer to the modern world's biggest question: How do we deal with the greater complexity of the "smaller world" created, in large measure, by the revolutions in transportation and in electronic communications?

Due to those two revolutions, we all "see" our own groups, other groups, and the combined consequences of all our actions quickly, in great detail, and with all the learned analysis we can stand — and more!

"Bad things" (e.g., pollution) once not widely noticed are now seen by larger numbers of people. For "activist idealists", this solves the previously gargantuan communication task of creating enough of a "shared view" to show that a problem exists. A constraint has been broken for them. Their basic motivation is increased by their perception that they now have a greater likelihood of success in fixing the "bad things". Add to this the perceived (and maybe real) increased scale of today's problems (although the Black Plague, for example, was a pretty big-scale event in its day). The result, for better or for worse, is that we have a greater number of potential activist idealists actually getting active.

A permanent condition, therefore, is created in which environmental and other activists will now always present decision-makers with more complex decision calculations. The smaller world has created a permanent condition of increased complexity of problems — more quagmires!

If an important characteristic of our times is that the times appear — or actually are — more complex, then our response should be to adopt tools that provide an improved ability to deal with complexity. Enter TOC. Quagmires call out for the TOC logic-tree thinking processes (TP).

## TOC Thinking Processes to the Rescue!

The personality type known as the "potential activist idealist" will most likely always be with us in this mixture we call humanity, as will its complementary type, the "status-quo and take-it-slow" (Edmund Burke style?) pragmatist/conservative type of person. So the challenge is to create processes that can draw out the best — and avoid the worst — of these two types of person, plus other types. In other words, these processes should:

1. Capture and value the inspired idea and/or idealistic vision (the trunk of the TOC future tree) from the activist idealists
2. Capture, value, and process the "negative branches" of the ideas (those aspects of the idea that lead to unintended, undesirable effects) from the "take-it-slow" pragmatists
3. Focus energy on identifying the constraining implicit assumptions in the planning by allowing the analytical types to articulate them

4.  Focus energy on developing additional ideas that further shape the idea (into its appropriate place in the "intrinsic order" or "natural solution" that is already there waiting for the project team to discover it) by turning the intuitives (the "idea people") loose on the problem.
5.  Work the problem back from the ideal outcome far enough to have a realistic and practical series of steps, however small, that constitute progress toward the ideal.

The TOC thinking processes look suspiciously like such processes.

## Activists Will Like the Thinking Processes

Will the activist idealists like the TP? Yes, because the future trees will help them share their vision more effectively (not to mention to refine and understand their vision or motivations better for themselves). Plus, the full TP suite will make them more effective as change agents.

## Conservatives Will Like the Thinking Processes

But what about the other side, the conservatives? I dare say that, were he alive today, even that archetypal conservative, Edmund Burke, would very much like the TOC thinking processes. His school of conservatism was not one of "make no progress" but rather (in TOC TP terms) to "make sure you perform the negative branch processing on your idea". Translating his classic example, the French Revolution's "rights of man" might have been a good idea, but a little negative-branch processing and transition planning might have created some of the same desirable effects without quite so many colorful and unnecessary undesirable effects.

# Obvious Fertile Fields for TOC Work

## Thinking and Health

What is the relationship between thinking and health? What ways of thinking give rise to health, well-being, and longevity?

These, in my estimation, are two of the central questions of our age. Actually, they have been among the central questions in every age. The difference — in our age — is that the answers are going to be made much clearer and will be acted on much more effectively.

How can I be so sure? Simple. The handwriting is on the wall. The evidence is becoming overwhelming. People such as Herbert Benson, Andrew Weil, Bill Moyers, Larry Dossey, Deepak Chopra, Norman Cousins, Christiane Northrup, Jeanne Achterberg, and Bernie Siegal (many of whom are doctors of traditional western medicine) are nudging the world in the direction of using the various forms of thinking more effectively for health. What results is a mix of thinking methods, self-care, nutrition, exercise, surgery, and drugs in a mixture and timing appropriate to each individual case. And why not?

Spiritual pioneers such as the Christian Science denomination, as well as some Eastern and so-called New Age groups will push the envelope of the faith and thinking portions of health, whether anyone asks them to or not. Science, though, needs to be there, gathering data, speculating causes, creating concepts, and characterizing the ranges of validity for simple and natural contributors to overall well-being. As with any other research area, TOC principles and structured thinking tools can focus and capture the thinking for individual lives or general public health solutions.

## Dr. Herbert Benson: Applying Science to Non-Mainstream Methods

I have to admit a favorite in this field, Dr. Herbert Benson. By publishing his 1975 book, *The Relaxation Response*,[4] Dr. Benson almost single-handedly legitimized a very simple, natural, and effective meditation technique by measuring and expressing its underlying cause-and-effect relationships using scientific methods and terms.[5] Meditation, of course, is a form of thinking. Dr. Herbert Benson has become a learned scholar of the placebo effect in the history of medicine, and of its potential today.[6] Two chapters in his book *Timeless Healing* are called "Faith Heals".

No need to attack the pharmaceutical and surgical industries. As Dr. Herbert Benson shows, health and well-being can be considered a three-legged stool, supported by pharmaceuticals, surgery and other medical procedures, and self-care.[7] As people make self-care more effective in their lives — including nutrition, exercise, stress management, and the various forms of thinking — their chances for the more serious forms of surgery and drugs are reduced.

## Simple Communication

To read the work of R. D. Laing is to know that his big heart must have been breaking to be able to see so accurately how awful human tragedies were being caused by simple failures to communicate and compounded by failures to understand and deal effectively with ever-present differences among modes of human experience. I wish Laing were alive to see the coming years of TOC logic trees being used by people — some of whom have literally been driven to distraction — to untie *Knots*.[8]

## Mental Health Support and Advocacy

I have an interest and some experience with issues in the mental health sciences and industry. I may have an opportunity to contribute there with TOC someday myself.[9]

## *Ecumenism*

I like this one a lot. Is there a tougher consensus-building problem? Church organizations will first use the TOC thinking processes as tools of administration. Then they will use them for strategic planning, which will lead them to policy. When church organizations begin to use the tools for policy, they will be ready to use them to forge greater areas of agreement and cooperation.

Are greater areas of agreement and cooperation needed? Let's look at a few numbers. There are approximately 6 billion people on Earth. Of these, approximately half — roughly two billion Christians of one sort or another, about a billion Muslims, and around, say, 20 million Jews – consider the Old, New, or both Testaments of the Judeo-Christian Bible to be important reference points. (Islam, of course, centers on the surahs of the Qu'ran which, in turn, acknowledge the Old and New Testaments of the Bible.) That leaves the other half — another three billion Hindus, Buddhists, Confucianists, Taoists, Shinto, Humanists, Atheists, and more — with other sources of primary beliefs, values, images, language patterns, traditions, and laws. In a world made small by communications and transportation revolutions, how will this dichotomy play out? During the past 20 centuries, one answer has been for one group to persuade the other groups to read their books and believe their beliefs. There may be limits to this now that did not exist before. This is, itself, a theological question that may take a generation or two to sort out.

Whatever you think about this, there is always the issue of what to do in the meantime. I believe the only way through this is for all groups to dig deeper to understand what they and their ancestors have been trying to accomplish with their books and beliefs. This opens the possibility of finding bases for cooperation that allow each group to:

1. Accomplish its most fundamental objectives
2. Continue to reading its own books
3. Accept that the others are going to be reading their own books, too

Digging deeper for shared purpose is one of the hallmarks of TOC and the logic-tree thinking processes. As more people find they like using TOC for projects at work and home, there is an increasing chance that people digging deep for shared purpose in the interfaith domain will be using the TOC thinking processes as one of their tools.

## *Environment*

### *Permanent Condition: No Turning Back*

Environmental matters are just one class of issues whose now-permanent presence in decisions makes managerial life more complex. TOC logic trees, informed by importance scales, can assist here.

# TOC and Information Systems

### *World Wide Web Becomes Worldwide Forest — of TOC Trees!*

This is a great topic. All of the arenas discussed above will move more quickly and have more impact due to the continuing remarkable improvements in the costs and effectiveness of information technologies. These changes continue to change the world in very fundamental ways.

The TOC thinking processes are very well suited to this ongoing information systems revolution. TOC logic trees, originally text and graphics, will be multi-media and hypertext knowledge frameworks with graphics, audio, video, programs, simulations, and hypertext linking tools. TOC logic-tree entities will be displayed on graphic screens with one-word or short summaries. Clicking on the entity symbol will cause an audio, video, graphic, text, or other information resource to be presented to the user. Entities will be linked via hypertext links to other entities and to supporting sources of information.

What cannnot help but emerge will be distributed compendia of TOC knowledge — about TOC principles and processes and about subjects that have been explored and characterized by TOC's self-documenting management-science, logic-tree thinking processes. As more people come to understand and use TOC, the logic trees will be built within and across distributed, global, information-sharing, power-sharing, and wide-area arrays of internets and intranets. This will allow large numbers of people to stay abreast of, and participate in, the analysis and discussion of important issues.

People will be assigned portions of TOC logic trees to prepare and maintain. Electronic mail and e-mail list servers will deliver opening, summary, and update messages to participants in discussions employing TOC logic trees. The messages delivered to participants will contain tree entities themselves built right in. Those entities will have hypertext links to the other tree entities — and other full TOC logic trees — prepared and maintained by others. The impact on the relative amounts of participatory (direct) and representative democracy, of this convergence of superior electronic telecommunications capability with superior structured thinking process technology, will be interesting to observe.

In other words, portions of today's World Wide Web will become worldwide forests of TOC logic "trees!"[10] The combination of the information technologies and the TOC thinking processes will cause the investments in information technology products, services, and training to pay off in spades, finally!

## TOC Project Management Systems

One of the important subsets of the intersection of TOC thinking processes and modern information systems technologies will be the automation of TOC project management tools. Project definition, planning, and implementation will be managed using software technologies. Thinking-tools modules will enable more rapid and efficient formulation of plans. Critical chain scheduling modules will assist with scheduling and rescheduling of resources throughout the enterprise, or across enterprises, to accomplish critical objectives.

## TOC Manufacturing Systems: De Facto Co-Standard

This one is well underway but not yet universally understood or robustly implemented. Several major systems vendors are taking a wait-and-see approach to the matter of giving customers a choice of traditional or public domain TOC systems modules. The argument has been made several times in the industry literature[11] that manufacturing companies, software companies, systems integrators, and other systems services providers are all well served by the trend toward having all important systems vendors providing APICS-supported public domain TOC information systems solutions. Here are the highlights of the argument:

- The manufacturing systems industry has a *de facto* co-standard.
- One half of the standard is based on the once arguably necessary assumption of infinite capacity in factory work centers. This is the traditional "closed-loop" materials requirements planning and capacity requirements planning (MRP/CRP) systems introduced in the 1960s. Though based on the implicit operational assumption of infinite capacity in factory work centers, MRP/CRP systems constituted a valuable advance over "order point", the predecessor technology they replaced, and continue to serve existing and new manufacturing company users well.
- The other half of the standard is based on the now inevitable assumption of finite capacity in factory work centers. TOC manufacturing systems are a new public domain generation of integrated manufacturing decision support, logistics control, and scheduling. They are based on the explicit assumption of finite capacity in factory work centers and constitute an important and very valuable advance over the closed-loop MRP/CRP systems.
- The TOC systems solutions are in the public domain and are supported by APICS educational resources.
- Both of these manufacturing computing alternatives can be used within modern manufacturing resource planning (MRP II), enterprise resource planning (ERP), just-in-time (JIT), lean manufacturing, agile manufacturing, and other higher level approaches to factory and supply chain management.
- An increasing number of manufacturing company customers are expressing a preference to have the TOC and finite capacity suite of tools included in systems company offerings. This allows companies to use TOC systems tools without extensive custom and semi-custom internal computer systems work.
- Software companies are strongly encouraged to support both the traditional closed-loop MRP/CRP and TOC approaches. Many systems companies are giving customers this choice, and all are encouraged to do so. When software companies support both approaches,

manufacturing companies have a choice between the two, offering a smooth migration path from the old to the new.

■ Integrated TOC systems of manufacturing decision support, logistics control, and finite capacity production scheduling are easier and more intuitive to use than the predecessor MRP/CRP systems.

■ As compared to — or in cooperation with — ERP, MRP II, JIT, lean, Toyota, Shingo, or other approaches, TOC strategies and systems deliver superior return-on-investment (ROI) because they focus on global (company-wide) vs. local (work center or departmental) performance, shrink investment in materials inventory to realistic minimum levels, maintain protective capacity at realistic minimum levels, and encourage priorities and decisions that support maintaining or increasing current and future cash flows (specifically, TVA).

■ As a result, TOC manufacturing systems are an inevitable industry standard. Eventually, the TOC systems will completely replace the traditional closed-loop MRP/CRP systems within MRP II environments. The higher levels of ERP and MRP II, such as sales and operations planning, demand management, and master scheduling will always be needed. They will remain with the new core TOC systems even after the lower level MRP/CRP systems are gone from the industry. In the meantime, it will take a long time before existing MRP/CRP systems, which are continuing to perform, are replaced with a new generation of systems. Also, in the meantime, many companies replacing the traditional system will continue to use MRP for ordering materials in a long-term horizon, even when it and CRP are removed from the system for decision support, planning, scheduling, supply chain control, and factory shop floor control. Eventually, constraints-based master planning processes will eliminate the need for the traditional MRP form of materials requirements planning systems.

■ All of which means that the higher-level ERP, MRP II, JIT, lean, Toyota, Shingo, and other processes and influences of manufacturing management will remain in place while the lower level, traditional, closed-loop, infinite-capacity MRP/CRP systems modules are being replaced with their TOC finite-capacity equivalents.

■ All of which means that the correct objective is to have TOC systems modules as standard selections on every important manufacturing software company's "menu".

### Beyond TOC Manufacturing Information Systems

Getting the TOC systems of decision support, shop control, and scheduling for factories on every important manufacturing software and services company's menu is the TOC systems equivalent of getting to first base in a baseball game. This section describes what amounts to getting to the other two bases, as well as hitting the home run.

- Manufacturing and supply chain software vendors are encouraged to include public domain TOC distribution and supply chain solutions in their enterprise and supply chain offerings. That's getting to second base.
- Next, management software vendors are encouraged to build public domain TOC critical chain project management scheduling features into all software that supports projects of any kind. That's getting to third base.
- Now for the home run. All software vendors are encouraged to include support for public domain TOC thinking process logic-tree processes and diagrams. If you think about what you have read in this book about the TOC logic-tree processes and diagrams, you will realize that they increase speed and effectiveness *everywhere*. Which means all software vendors of all applications can do themselves a favor by building homes for TOC trees in their offerings.

It is all in the public domain. Large numbers of people and companies are getting acquainted with TOC and are going to want to have elegantly executed and well-integrated information systems support for all these TOC innovations. They will not want to provide the systems support via their own internal custom or semi-custom programming, unless the industry, through inaction, forces them to do it for themselves. To all software vendors and systems services providers, don't say I didn't try to tell you, and good luck!

# TOC for Education

### Middle School Children Show Industry How It Is Done

I watched in amazement as one young child after another stepped up to the overhead projector at the June 1994 TOC conference in Florida to present their real-life applications of the TOC thinking processes. Never again have I taken seriously any adult's comment that the TOC thinking processes are "too complicated" for them to understand or use. And no more did I wonder about whether this latest round of innovations from Dr. Goldratt was going to have an impact, not only in management practice in industry, but also in the quality of day-to-day life. These young people described how they increased their study efficiency, improved their relationships with family members, and helped friends or themselves step away from destructive paths back to constructive directions. I can tell you that, for some of the more touching presentations, there was not a dry eye in the conference center.

The adult presenters included professionals within the Okaloosa County School System, including Kathy Suerken, a seventh-grade teacher, who later takes on a larger role. During the presentations, Kathy Suerken gave each attendee a copy of her letter of June 1992 to Dr. Goldratt which had gotten it all started. A teacher at the Ruckel Middle School in Florida, Kathy had used *The Goal* in connection with her work as a volunteer leader of International Projects for the school and in a World Cultures course. *The Goal* had found its way to the middle school from a program office at nearby Eglin Air Force Base in one of those series of events that makes fiction pale in comparison to real life. The letter's purpose was to hint that the school and the Goldratt Institute might collaborate in developing a course called "Learning to Learn (Think, Lead)". In the conclusion of her letter, Kathy speculated that there may be a convergence between the goals emerging in the middle school's programs and Dr. Goldratt's own personal goal. She closed the clever letter with "only time — and you — can tell me if that is so."[12]

Well, it *was* so. She and the school got their answer in spades. Dr. Goldratt sent Kevin Fox, one of his young Odyssey Course graduates, to Florida to work with the teachers and administrators. Later, Goldratt Institute partner, Dale Houle, worked with the group. The presentation I saw in June 1994 was one of the results. More would soon follow.

## TOC for Education, Inc.

Shortly thereafter, Kathy Suerken formed a non-profit foundation, TOC for Education, Inc., for the purpose of disseminating the logic-based tools and common sense methodologies of the Theory of Constraints into K–12 (kindergarten through twelfth grade) education systems worldwide. Dr. Goldratt and his institute continue to donate their knowledge and resources to the effort.[13]

The professionals at TOC for Education use the TOC tools at several levels. They teach TOC outright to educators and support use of the tools by children, but they also use the thinking processes for their own operations planning.

They used the thinking processes to create and implement the strategic and tactical plans for the foundation. After just 15 months, they had developed programs in ten states and four foreign countries — programs that are beginning to have a significant impact on student behavior and learning. Kathy Suerken provided the following update.[14]

## High School's Peer Mediation Program

One of the most powerful and immediate applications of TOC to educational settings is in conflict resolution. After completing a TOC for Education workshop in August 1996, Assistant Principal Ben Walker of South Gross Pointe High School in Michigan wrote, "I plan to introduce the entire concept into our school's Peer Mediation Program."

Demonstrating what can be accomplished when dedicated, talented people think and act on the causalities between the teaching, learning, and use of TOC, Ben taught the evaporating-cloud and negative branch processes to a group of students. He then supported them in their work to adapt the program to TOC principles. In less than 3 months, not only has the revised program been formalized as a handbook, but it is "being successfully implemented every day," according to team leader Will Stentz, a high school senior who gave up football to ensure the substantive, compelling results from the school's new Advanced Peer Mediation Program.

"The entire team feels this is the most effective style of mediation we have ever used. Our previous program only addressed people's wants, which led to compromise. Conflicts are a part of life — the way we handle those conflicts makes the difference. What we have at South is the future of conflict resolution — a program that targets people's needs."

Like most people who become involved in TOC, Will's personal goals are becoming more global. "The only way to have a peaceful community is to spread the knowledge," he says. An indication of their overwhelming success with TOC was the invitation to Will and eight other students to spread their knowledge at the 1997 annual conference of the North Central Association Commission on Schools. This commission provides accreditation to schools in 19 states, to selected Department of Defense dependents' schools, and to the schools of the Navajo nation.

## Clouds for Everywhere?

The propagation of tools that work is also the desired outcome of his high school peers in Florida. "This idea, the 'cloud', is a useful tool that should be used by everyone to solve everyday problems or problems that could completely change our future. The cloud, in my opinion, is the best problem-solving technique that has ever been invented and should be shared with her peers everywhere," writes Shane Moore of Niceville High School, FL.

## Preventing Conflicts

In addition to resolving conflicts, TOC is being used in the classroom to prevent them. This application focuses on enabling students to see the logical connections between what they do and what happens to them as a consequence of their own actions so that they can begin to make more responsible choices. Enabling children to think things through systematically can change their perspective: "If you look at my tree, you can see the letter is a S. It's a S because Steven was my enimie then. Now I think we become friends," observes sixth-grade East Los Angeles student, Rachell Alcarez.

## Teacher's Role: To Teach Students How To Think

Benefits such as these impact the entire classroom, including teachers. "I have been searching for an idea to teach children on how to be more peaceful. I went to graduate school at Notre Dame in International Peace Studies. I discovered peer mediation and came here to use that tool to help create more peace in L.A. I feel the TOC process to be the most powerful tool that I have ever seen to help people create personal, interpersonal, and ultimately global peace," writes California science teacher Geri Portnoy. "I am glad, too," she adds, "to know I now have a tool to teach children how to think. In this way, I feel like a real educator."

## Using TOC To Design — and Within — Curricula

A thinking curriculum is the brass ring to educators such as David Kibbey, Curriculum Director, Area F, Detroit Public Schools. "I will use my TOC knowledge with schools and school improvement teams to help them design the best, most productive plans to raise student achievement. We will spread TOC into classrooms throughout Areas F and D by training as many teachers and counselors as our time, resources, and strength will allow. You have ignited a fire in southeastern Michigan. May it burn out of control and consume the apathy, cynicism, and dismay that infect us."

The application of TOC to curriculum begins with clear lesson plans that ensure that classroom activities actually lead to curriculum objectives. TOC can be taught outright as critical thinking skills and/or synthesized with existing curriculum. The process of systematically thinking things through is as applicable to an English composition lesson as it is to a science fair project.

## Cause-and-Effect Thinking Delivers Better Student Essays

"I have taught research-based papers for many years," writes high school English teacher Beverly Brown of Ashland, Ohio. "For years, students have had difficulty with the body — organizing ideas and maintaining unity. After our discussion of cause-and-effect, I realized the students need to understand this concept before we write the essay. I will now work on cause-and-effect before I teach essay writing. This course has changed my entire way of teaching. I now see the process is the problem, not the student."

## Establishing Respect for Teacher in East L.A. Classroom

Referenced in Rotary International's *Peace in the Community* publication, Niceville High School student Becky Barr thoughtfully echoes that sentiment: "TOC enables others to realize that it is not the people that are the problem; it is the situation." Marcia Hutchinson, a counselor at a middle school in East Los Angeles, did not stop there. She used TOC to take the changing perspectives of an entire class to a whole new level of respect and dynamic of learning in a classroom that had actually become abusive to the teacher.

## Elementary School Principal Says TOC Is a Foundation for Change

Bob Bohnstengel, a South Carolina principal, changed the dynamics of an entire school culture with a systemic application of TOC: "I truly believe TOC to be the foundation for change at my school. The Ridgeland Elementary School faculty has identified the need for better communications, a clearer sense of teamwork within the faculty, and empowerment. All of these needs are addressed through the TOC thinking process," which he is bringing to his entire faculty and staff by personally training them after hours.

## Students Plan Together

Noting improvements in communication at all functions, Bob Bohnstengel also refers to the synergistic results from a curriculum application: "Recently, a second-grade class incorporated their TOC skills into a science lesson to plan a flower garden at the school. When they explained to me the process they had used, I thought that not only are these second graders really thinking, but they are actually working together as a team to solve problems."

The class of seventh-grade students at Ruckel Middle School in Niceville, FL, set out on a journey of "creating a better and more thoughtful world". One of the students, Jesse Hanson, expressed a more specific wish: to ensure that the "TOC techniques were shared with kids who have a tougher set of problems."

## Impact Within Juvenile Detention Camp

Following a 1996 experiential demonstration of TOC with a class of 15- to 18-year-old students in a juvenile detention camp, a student said, "I admit I was feeling kind of hopeless about making it. Even making it to 21 was hard to see in my future. But you gave me hope and the thought that I can change and that I can make it if I just try and follow those steps."

And from other student: "Well, first of all I thought the lady was making a lot of sense. I hope she succed's [sic] because if she does it will benefit young people. It will help keep people from coming to jail and just help people make the right choices and make something positive out of their lives." It would appear that Jesse Hanson's wish is coming true.

## Harnessing Emotion and Putting It to Constructive Use

Kathy Suerken summed it up: "TOC enables children and adults to harness their emotion, to make it work for them in a positive way. As a result, they are better able to clarify and articulate their thoughts, leading to vastly improved communication and relationships with others — along with the realistic self-confidence which comes from knowing that they are capable of solving their own problems. In other words, when people practice TOC, they develop a sense of self-empowerment.

"The TOC skills provided through TOC for Education workshops unleash children's potential to take responsibility not only for their own learning, but also for their own behavior. The word educate, after all, is derived from the Latin word *educare*, which means to lead out.

"Why do I like TOC? TOC truly educates and, in the process of so doing, enables people of all ages to bring out the best in themselves — a 'best' which keeps improving. Where do you think *that* will lead?"

Time will tell. To help people keep track of Kathy's progress, I have established a web page at http://www.tbmcm.com/toc4educ.htm that will provide links to the TOC for Education's pages.

## Avery-Dennison Manager Takes Aim at University Education

At a break during the 1996 APICS Constraints Management Symposium, a young man approached me in the bookstore and asked me why TOC wasn't taught in more business schools. I said we are working on it and asked why. Cary Glay, Cost and Manufacturing Analyst for the Specialty Tape Division of the Avery-Dennison company, tells this story: "When I reported for work, fresh out of business school, the first thing the senior financial manager asked me was whether I had read *The Goal.* When I said no, he told me to forget everything I'd learned, read the book, and come back to discuss it with him. I did. I'm very glad the company has been giving me the time to get up to speed on TOC, but I'm angry that my professors had given me no clue that this body of knowledge existed. Not even a clue! And the methods are right. So, what I want to do is give every school the chance to add TOC to their curriculum." Cary is working on this. To help people find him and keep track of his progress, I have established a web page at http://www.tbmcm.com/toc4coll.htm.

## Harvard Business School: Logic Trees for the Case Method?

I have not told my *alma mater* yet, but I think first-year Harvard Business School students should be encouraged to use TOC logic-tree thinking processes to prepare, present, and discuss their famous three case studies per evening. Hey, I'm entitled to an opinion!

## U.S. Naval Academy: Logic Trees for Policy and Strategy?

I have not told my *alma mater* yet, but I think Naval Academy midshipmen should be encouraged to use TOC logic-tree thinking processes to practice formulating policy, to analyze the evolution of technology and techniques described in *Seapower*, and to prepare tactical combinations. Hey, I'm entitled to an opinion! (Why does that sound familiar?)

## People Who Rely on Formal Education Will Fall Behind

TOC makes it easier for people who already have the skills that come from succeeding without formal education to develop the most important core of the skills and perspectives which normally only result from the better programs of traditional formal education.

Think about that.

## Just Do It!

"Practice makes perfect."
> ∾ Moms and dads everywhere

"Do or do not. There is no try."
> ∾ Yoda[1]

"Just do it."
> ∾ Nike


## Knowing What You Want Can Be Half the Battle

"When Siddhartha has ... a goal ... he goes through the affairs of the world like the stone through the water."
> ∾ Siddhartha to Kamala, in *Siddhartha*, by Hermann Hesse

"Be careful, Richard."
> ∾ Don Shimoda to Richard Bach, in *Illusions*
> (the "Blue Feather" scene)

"Know thyself."
> ∾ Socrates, according to people said to have known him


## Every Day Is a New Day

"Every generation yields
The new born hope
Unjaded by their years."
> ∾ Sarah McLachlan, "Wait", *Fumbling Towards Ecstasy*

# 6 How Do I Get Started With TOC?

This chapter (1) provides general advice and five steps for effectively introducing the TOC management system within an organization; (2) suggests seven things the reader can do right away to get started with TOC, even if his or her organization has not yet decided to introduce TOC as part of its foundation in management principles and practice; and (3) cites several current sources of more detailed implementation advice, while indicating that more are on the way.

## Five Steps To Introduce the TOC Management System

Your company has decided to introduce and use the TOC management system as an important portion of its thinking and practice. Over a period ranging from 6 to 18 months, the management will guide the company through the following five steps:

1. *Conduct TOC education.* Introduce the basic TOC principles and processes through a program of education and training. It will normally be useful also to conduct additional education for specific needs (e.g., in support of initial applications of TOC logic trees to strategy, process improvement, and project management; building TVA financial management systems; and incorporating TOC principles into factory and supply chain applications).

2. *Integrate TOC with other management approaches.* Facilitate the integration of TOC with the company's other existing management and improvement approaches.

3. *Build a TVA financial management system.* Develop an internal TOC financial management system centered on maintenance and growth of the TOC throughput value added (TVA) money flow as a perpetual top priority.

4. *Complete TOC projects other than major changes to logistics infrastructure.* Encourage the early and sustained use of TOC principles and structured thinking processes (logic trees) for specific projects in the areas of strategic positioning, operational policy, and process improvement.

5. *Establish TOC foundations for logistics.* On the strength of the above foundation, support the introduction of drum-buffer-rope factory scheduling, buffer management logistics control, and (where appropriate) TOC distribution, supply chain, and critical chain processes in support of resource and capacity management.

## *Preliminary Words of Advice*

Always begin with education; the many ways to do this are covered in the next section. Make sure what you are trying to accomplish with introducing TOC makes sense. I sometimes hear the comment, "I just want to implement drum-buffer-rope scheduling because I've heard it's the best scheduling technique." This only makes sense if the thinking and management accounting in the organization are going to change too. The same goes for wanting to use TOC or drum-buffer-rope to cause a quick hit in profits, inventory reductions, or lead-time reductions. To try to go too quickly is, in most cases, to go more slowly or to never get there at all.

In other words – even and especially when you are in a hurry to implement drum-buffer-rope — it is best to implement TOC methodically and correctly. This gives you support for "can-do" attitudes, improved performance measures, greater ability to change policies smoothly, faster introduction of new products, impetus for increasing sales and TVA, impetus to enter new markets — all prior and in addition to — implementing drum-buffer-rope scheduling, buffer management shop floor control, or other major changes in logistics. This sequence removes many sources of confusion during implementation, provides new tools and skills for dealing with the policy changes required for the larger drum-buffer-rope and other logistics projects, and places cornerstones in the foundation for your TOC management system.

Always start as soon as possible with the TOC thinking processes, the logic trees. Where to start? The way to start, when you do not know where or how to start, is to begin with TOC current-reality trees using the initial list of 6 to 12 undesirable effects. This process is illustrated in the *Mission Impossible* example in Chapter 1, in *It's Not Luck,* and in William Dettmer's 1997 thinking processes book, among other places. This will usually lead to ideas, or vision statements, that can be worked through using the TOC future logic trees. The essence of the logic trees lies in their individual facts or entities, the individual cause-and-effect links, and the thinking and cooperative disciplines of the categories of legitimate reservations (CLR). The simple discipline of nailing down one "if … then" statement after another is where the action is. Do not make it complicated. It may be difficult (or easy), but only to the extent that any rigorous (and creative) thinking is difficult (or easy), but it is *not* complicated. This difference makes all the difference. Start early with the TOC logic trees. Start now.

## Lessons of Experience

The Theory of Constraints has been introduced into organizations in many ways. Some "good" ways recommended by experts have been failures; some "bad" ways have been successful. There has been everything in-between. Many of the past approaches to TOC implementation were born of necessity due to the lack of tools and lack of management and industry consensus on TOC.

## Objectives of the Implementation Process

The approach recommended in this book is, in my judgment, the best approach, given the experience to date and given the tools available today. Compared to other approaches, this one:

- Minimizes expenditures for products and professional services
- Delivers the most, or the broadest array of, important benefits
- Increases the likelihood of success
- Reduces or eliminates the known risks

## How Much TOC, at What Level of Detail, Is Enough?

Determining appropriate levels of detail for implementing TOC solutions is part of the planning process. Small companies, or large companies with very simple operational characteristics, can use simpler versions of TOC solutions. Larger companies, or smaller companies with high operational complexity, can use the same TOC solutions but in their more comprehensive and detailed forms.

In almost all cases, even where a more detailed and comprehensive form of a TOC solution will eventually be desired or needed, the company can begin with the simpler form and enjoy gaining the larger share of that solution's benefit. As incremental benefits, controls, or other features are then desired, the next level of detail in the solution can be introduced.

For example, while all manufacturing companies will use at least simple drum-buffer-rope scheduling methods, not all will need to invest immediately in the additional understanding and tools required to use the scheduling, schedule evaluation, and capacity management methods based on dynamic buffering. Some companies will be simple enough that they will never need dynamic buffering at all.

All TOC companies will calculate and use past, current, and future TOC throughput value added (TVA) as their focal points for financial and operational planning and control. This refers to entities 16a and 17a on Figure 1.1. However, not all companies will find it immediately worthwhile to take the calculations beyond TVA and operating expense (OE) to produce a full set of balance sheets and income statements based on TOC. At some point, the financial computing systems industry will remove this issue by providing them as standard reporting formats.

All companies will use the basic TOC principles of system, goal, physical and policy constraints, global and local measurements, importance scales, TOC throughput value added (TVA), three management roles, five focusing steps, structured logic-tree thinking processes, and project management. These are the TOC elements shown as entities 5 to 7a, 11, and 13 to 17a in Figure 1.1.

All companies will use TOC logic-tree thinking processes to guide and implement projects (entities 7a and 11). Fewer companies will have projects with the complexity or scale to merit the critical chain project scheduling technology (entity 18a).

Starting early — to overcome procrastination in getting the first several simple and then important projects accomplished with TOC logic-tree thinking processes — ensures that the company will establish and maintain habits of accelerated and self-documenting processes of project management, organizational learning, and continuous improvement.

After introducing TOC concepts and processes that are applicable to all organizations, a team can consider identifying, adapting, and gaining skill with selected TOC solutions that apply only to specific industries or situations. For example, a manufacturing company that has introduced the basic TOC management system and then derived great benefits from basic drum-buffer-rope and buffer management may consider introducing more advanced forms of TOC manufacturing scheduling and shop floor control (entity 18a).

After the initial introduction of the TOC management system, the company continues to use TOC to deepen its expertise within its specific business knowledge areas and to increase its skills with TOC itself.

## TOC Educators and Consultants

Here are a few things to consider concerning use of TOC educators and consultants. The philosophy underlying the TOC management science is to get companies out from under the effects of management fads and to empower the employees of the company (vs. external consultants). This is a fine baseline. It is healthy. Companies should do as much internally for themselves as possible.

Still, there are situations where external parties are useful. History, employee composition, or timing can all dictate that it is best to have a consultant-supported, or even consultant-led, TOC implementation.

# Step One: TOC Education

Here are three suggestions for conducting basic TOC education:

1. Make effective use of the TOC novels, especially *The Goal* and *It's Not Luck.*
2. Provide and encourage use of a TOC resource library.
3. Conduct TOC education and training.

## Use the Primary TOC Business Novels Well

The most effective and inexpensive way to conduct company-wide TOC education is to encourage people to read *The Goal* and *It's Not Luck.* Not every company has the product and plant structure, policy and information systems infrastructure, or organization culture that allows creating excellent improvements in performance simply from having some right group of people read and study these two books. On the other hand, many companies *have* had such circumstances and have achieved the desired results. My suggestion is have everyone who is considered (or who wants to be considered) a serious player in the company's TOC initiatives read both basic TOC books twice and well. If the company can go to the bank laughing right from there, fine. If not, the team is in a better position to employ more deliberate approaches for more complicated situations.

Encouraging people to read *The Goal* — including the two introductions contained in the 1992 second revised edition — sends the messages that people are encouraged to think. They are encouraged to be physicists within their organizations and to challenge current practice in a constructive and practical manner. They are encouraged to use these techniques and habits to focus the organization's resources on creating cash flows from and for continuing operations — making money, increasing TVA — as the perpetual imperative. *The Goal* uses role models in the single-factory setting, at work and at home.

Encouraging people to read *It's Not Luck* presents effective role models at the level of a corporate board of directors and associated senior executive staff. It provides an introduction to the structured TOC logic-tree thinking processes and provides exposure to several valuable solutions in operations and strategy. Several chapters show examples of how to begin in the classic situation for which it is not otherwise clear where or how to begin (i.e., the "6 to 12 undesirable effects" approach to building a current-reality logic tree). Several evaporating-cloud diagrams are applied to realistic work and home situations. Chapter 26 of *It's Not Luck* describes the steps for a prerequisites tree in the context of career planning. The transition logic tree is presented in one of its most valuable contexts, the development and fine-tuning of direct selling and negotiating methods in support of breakthrough tactical and strategic marketing solutions. The preference to grow struggling businesses into valuable assets, rather than sell them at distressed prices and trigger layoffs, is articulated along with motivational images of success. The explicit balancing of customer, shareholder, and employee interests is addressed in the context of the effects on strategic planning. Adding *Critical Chain* to the encouraged reading list adds an introduction to TOC project management, scheduling, and measurement concepts.

The question often arises: Which of these three TOC novels to read first? In my estimation, it does not matter. If you are pressed for advice, then people in manufacturing and other operations should start with *The Goal*, then *It's Not Luck*, and then *Critical Chain*. Directors, executives, and managers outside operations should read in the sequence of *It's Not Luck, The Goal,* then *Critical Chain.* When in doubt, read *The Goal* and *It's Not Luck* twice before moving on to *Critical Chain* or other TOC books. Speaking of other TOC books and resources, let's take a look at some of what is available at this writing.

## Build the Company TOC Resource Library

APICS keeps abreast of all the TOC literature and carries all of the best. At this writing, due to the efforts of the APICS Bookstore staff led by Michael Clark, Paula Smith, and David Strickland, APICS has the world's finest and most comprehensive selection of TOC books, article reprints, and audiotapes.

## APICS TOC Conference Symposium Proceedings and Audiotapes

Chapter 4 contains a list of most of the APICS Constraints Management Symposium proceedings and audiotapes (with their individual APICS stock numbers) from the years 1995 to 1997. The APICS Constraints Management symposia were designed and implemented with great care. They provide an affordable, high-quality, comprehensive, and authoritative source of TOC concept, technique, and case study information. The information applies across a wide range of company sizes, organization types, departmental views, and industries. There are also a few tapes (e.g., Dr. Goldratt's 1995 presentation) which provide historical perspectives.

Don't make the mistake of underestimating the enormous practical value of this collection of APICS Constraints Management Symposium proceedings and audiotapes just because they are not very expensive!

Each of the audiotapes can serve a self-education and perspective-building function for your company. Many of the individual presentations are referred to in this book, both in the notes and especially in Chapters 3 and 4. Due to their high value and low price, the correct strategy is to make the full sets of all the available written proceedings and audiotapes available within your organization. At this writing, the 1995, 1996, and 1997 sets are available.

## TOC Publications

The following are many of the TOC publications available from APICS. The APICS stock numbers are shown with each selection.

*The Goal: A Process of Ongoing Improvement,* by Eliyahu M. Goldratt and Jeff Cox (1992, item #03201). The advice here is simple. Everyone should read it at least once carefully, including and especially the two introductions. Many should read it twice, including introductions. People leading TOC work in logistics should read it twice within a few months, and then once again after about a year.

*It's Not Luck: The Story of Alex Rogo,* by Eliyahu M. Goldratt (1994, item #03291). Same advice as for *The Goal.* This one is for everyone, too. People leading any work with TOC logic tree thinking tools should read this twice within a few months, and then every once in a while.

*Critical Chain,* by Eliyahu M. Goldratt (1997, item #03203). This third business novel from Goldratt concerns the TOC approach to project management. It's good. It's useful. People new to TOC can go a long way with *The Goal* and *It's Not Luck,* and then add in *Critical Chain* when they are good and ready.

*The Haystack Syndrome: Sifting Information from the Data Ocean,* by Eliyahu M. Goldratt (1990, item #03125). *The Haystack Syndrome* describes the simple, inevitable, and comprehensive replacement for allocation-based product costing for management accounting and decisions. It develops the important theme of moving from a "cost world" to the "throughput (TVA) world". It documents the well-known "P&Q" decision exercise that — while illustrating product planning based on TVA per constraint unit — describes the problems with traditional cost accounting and highlights "inertia in thinking" as a frequent root cause of industry's problems. *The Haystack Syndrome* provides a public-domain specification for TOC decision support, shop control, and finite-capacity factory scheduling. This book is *not* a novel and is a tougher read than *The Goal* and *It's Not Luck.* Part One is about the changing decision process and is useful reading for all managers. Part Two is useful for everyone involved in implementing TOC logistics implementations. Part Three is necessary reading for all developers of TOC manufacturing computing systems, TOC master schedulers, and all leaders of implementations of TOC *The Haystack Syndrome* logistics software systems.

*What Is This Thing Called the Theory of Constraints and How Should It Be Implemented?,* by Eliyahu M. Goldratt (1990, item #03341). This book was written in a hurry, in the midst of several other major projects during 1990. That it was written at all during that busy year is amazing. As to its strengths, the value of the book lies in the fact that Dr. Goldratt explains — in terms of generic cause-and-effect, and in terms of his organization's vast implementation experience — why the products and services are designed and used in certain ways. Therefore, an implementing company can gain an understanding of some important implementation dynamics, and a consulting firm can learn how to design elements of its own TOC practice. It is worth reading, but not before *The Goal* and *It's Not Luck* are each read twice. Since I am recommending the book, I should also caution as follows: The rush to complete the book shows in the heavy use of excerpts from other Goldratt publications, typographical errors, and an occasional lack of coherence. More importantly, though, it answers the questions contained in its title in terms of the specific services and products available only from the Goldratt Institute at that time. (Readers normally expect generic issues and know-how from books, which they will make, buy, or use as they see fit; specific products and services are normally the province of marketing brochures.) Some of the specific advice is out of date since the TOC knowledge base has continued to evolve. The book fails to acknowledge the other TOC educators, consultants, and academics who do the same things in different ways. If you are leading TOC work, you should read it.

*The Race*, by Eliyahu M. Goldratt and Robert E. Fox (1986, item #03202). This book has the format of overhead slides on odd-numbered pages with notes and comments on the facing even-numbered pages. Many companies have used the slides from *The Race* as materials for internal workshops. It is not a novel and not a Socratic book. The book contains an excellent graphical analysis of the effects of high inventory on throughput value added (TVA), inventory, and operating expense. Several terms, including synchronized manufacturing, drum-buffer-rope, buffer management, time buffers, and holes in buffers are introduced. The linkage between drum-buffer-rope and improvement processes is articulated. The picture of Goldratt and Fox smiling together with their trademark fine cigars is worth the small list price of the book (at this writing, about $15.00). It made some of the solutions in *The Goal* more clear, but other books that came later did that better — for instance, the next three items on the list from Dr. M. L. Srikanth and his colleagues at Spectrum Management and Dr. Michael Umble of Baylor University.

*Regaining Competitiveness: Putting* The Goal *to Work*, by M. L. Srikanth and Harold Cavallaro (1993, item #03931). This book does what its title says it does — gives concrete advice on how to put the approach introduced in *The Goal* to work.

*Synchronous Manufacturing*, by M. L. Srikanth and Michael Umble (1990, item #03602). Readable conceptual depth on drum-buffer-rope and buffer management and a high-level overview of TOC measurements. There is also a companion workbook (1992, item #03357).

*Measurements for Effective Decision-Making,* by M. L. Srikanth and Scott Robertson (1995, item #03355). Blends TOC measurements with the format of the balanced scorecard.

*TOC and Its Implications for Management Accounting,* by Eric Noreen, Debra Smith, and James Mackey (1995, item #03356, with field case studies). This landmark book is discussed in Chapter 2. Some of its case studies are summarized in Chapter 4. This is a very useful book to provide to financial managers and executives. One of its appendices contains a good overview of the TOC logic-tree thinking processes. Its main chapters contain quite a few sample logic trees from actual industrial situations.

*The Theory of Constraints (TOC): Applications in Quality and Manufacturing,* by Robert Stein (second edition, 1997, item #03338). This book provides concept and detail for how TOC is applied to make Total Quality Management more effective, primarily in manufacturing companies. It discusses using focus to increase profitability and to avoid burying people with too many projects (causing nothing to get done well). TOC is described in conjunction with statistical process control (SPC), design of experiments (DOE), and other important Total Quality tools. The title of this book was once *The Next Phase of Total Quality: TQM II and the Focus on Profitability* (first edition, published in 1994). I had suggested to the author that he rename the book to include TOC. Some changes in content may also have been made.

*Reengineering the Manufacturing System: Applying the Theory of Constraints,* by Robert Stein (1996, item #03480). I wrote the foreword for this book. It is a very good follow-up to *The Haystack Syndrome,* is written in a style more typical for such books, adds data structure and implementation issues beyond the intended scope of *The Haystack Syndrome,* and adds the TOC logic trees that had not been invented at the time of *The Haystack Syndrome's* publication in 1990. This book moves *The Haystack Syndrome's* TOC computing a strong step forward. My next book in the APICS/St. Lucie Press TOC Series will take that same subject — public-domain *The Haystack Syndrome* TOC manufacturing systems — the several additional steps required for all the necessary information to be in one place (vs. scattered around the literature and the "oral" public domain).

*Goldratt's Theory of Constraints: A Systems Approach to Continuous Improvement,* by William Dettmer (a.k.a. "Dettmer's book about TOC logic-tree thinking trocesses", 1997, item #03516). Contrary to the impression given by the title, this is not an introduction to TOC, but a detailed textbook on the

TOC logic trees. I wrote the back cover praise for the book as follows: "If you are an experienced TOC implementer, read Bill Dettmer's book to deepen your understanding, increase your skills, and accelerate the pace of your TOC projects. If you're new to TOC, fasten your seatbelt now; you'll soon be making a more fundamental and more constructive difference — on a greater number of more important and complex projects — than ever before." At this writing, this is still the best book for gaining detailed information and step-by-step procedures concerning the TOC thinking processes and the logic trees. Another such reference book is expected soon from TOC logic-trees expert, Lisa Scheinkopf, as part of the APICS/St. Lucie Press TOC Series.

*APICS Selected Readings in Constraints Management* (item #05021). This is an APICS publication with a excellent selection of papers from Goldratt, the present author, and others on a wide range of TOC topics. Goldratt's landmark 1983 paper, "Cost Accounting: Public Enemy Number One", is included. Also included is an article I wrote about using a co-standard of manufacturing systems, "The Systems Industry Is Giving MRP (and ERP) Implementers a Choice: Traditional and TOC Systems". This paper contains historical, technical, and diplomatic sections to demonstrate that there is good reason that an increasing number of manufacturing systems companies are providing both the traditional closed-loop MRP/CRP and the *The Haystack Syndrome*-influenced TOC solutions as well-integrated standard selections on their menus of modules. There are also selections concerning TOC and agile manufacturing, new product introduction, purchasing, new plant design, QS-9000, integration with just-in-time, accounting, project management, and both case study and concept material concerning finite capacity factory scheduling.

*Constraints Management Handbook,* by James Cox and Michael Spencer (1997, item #03522). This is the Jim Cox who has created a large network of doctoral and other graduate students who are widening and deepening the TOC body of knowledge and experience in universities and organizations around the world. At this writing, I have not seen the book, but it no doubt is very useful. This is a book in the APICS/St. Lucie Press TOC Series.

*Tough Fabric: The Domestic Apparel and Textile Supply Chain Regain Market Share,* by John W. Covington (1996, item #03436). This tells the story of how John worked with Eli Goldratt on the pioneering case study described in the book's title. Prior to the TOC work, a computer system was thought to be the answer to the industry's problems. It wasn't, of course. *Thinking* helped to sort out what the computer could, could not, and should do. Another book about TOC applications to supply chains by the same author is expected soon in the APICS/St. Lucie Press TOC Series.

*Project Management in the Fast Lane: Applying the Theory of Constraints (TOC),* by Robert Newbold. Rob knows of what he writes. He did much of the critical chain project management software development work for Dr. Goldratt's initiatives to demonstrate concepts. He has created some software of his own (compatible with Microsoft Project). This is a book in the APICS/ St. Lucie Press TOC Series.

*Securing the Future: Applying the Theory of Constraints,* by Gerald Kendall. This book in the APICS/St. Lucie Press TOC Series shows TOC applied to strategic planning and implementation, complete with sets of sample logic trees for cast studies.

Other TOC publications — new titles are added frequently. For the current list of TOC and other educational resources from APICS, request the current APICS Educational Materials Catalog. The catalog is free from APICS as item #01041. The catalog and all of its selections are available from APICS Customer Service at (800) 444-2742, ext. 2350; (703) 237-8344; fax (703) 237-1087, and Internet http://www.apics.org.

## APICS TOC Resources Available via APICS Online

The Internet site managed by APICS is an excellent resource for information about TOC and other topics in resource management. At this writing, the APICS site on the World Wide Web is located at http://www.apics.org.[2]

## TOC on the Internet

Internet resources have this maddening habit of moving from one electronic Internet address to another. Some disappear entirely for reasons known only to the owner of the sites. APICS maintains a list of useful sites and discussions in various knowledge areas, including TOC, at http://www.apics.org. I keep track of TOC sites and make hypertext links to them from the McMullen Associates web site at http://www.tbmcm.com. Another way to locate TOC information on the Internet is to use Internet search engines,[3] using phrases such as "Theory of Constraints", "TOC", or "Constraints Management".

## Conduct TOC Education and Training

There are many providers of TOC education and training in today's marketplace. Many of them can be encountered via education and networking at

APICS. In addition, APICS itself provides TOC education in both open public and dedicated in-house settings.

## *Join APICS and the APICS Constraints Management SIG*

Any company taking TOC seriously would do well to have some of its employees become APICS members and join the APICS Constraints Management SIG. This will place them on a distribution list for all the information provided to members, not only about TOC, but also about the wide range of other education and certification resources APICS provides.

## *Use APICS Constraints Management (TOC) Symposia*

The proceedings and audiotapes of these events have been described. People should, of course, also attend the events to observe the presentations first-hand, meet the presenters and exhibitors, and network with fellow attendees. At this writing, the annual Constraints Management Symposium is held in April of each year.

## *Use the TOC Portion of the APICS International Conference*

The annual APICS conference provides many TOC presentations and exhibitors as part of a much larger event. At this writing, this event is held every October.

## *Use the APICS Theory of Constraints Workshops*

The APICS education department and the APICS Constraints Management SIG have developed a family of APICS TOC Basics Workshops, as well as courses on advanced topics. At this writing, these courses include:

- Introduction to the TOC Management System
- TOC Logic-Tree Thinking Processes
- TOC Measurements (TVA Financial Management System)
- TOC Production Management
- TOC Supply Chain Management
- TOC Project Management
- TOC for Financial Managers

- TOC Manufacturing Systems
- TOC for Quality and Manufacturing

## Make Use of the APICS Theory of Constraints Certifications

Currently, TOC knowledge areas are included in portions of the APICS internationally respected certification programs, Certified in Integrated Resource Management (CIRM) and Certified in Production and Inventory Management (CPIM). Additional TOC information is being added to these well-known programs. In addition, selected separate APICS certifications are under development for specific industry and application market niches.

## Use the TOC Logistics Games

There are three classic TOC educational game families that are useful for teaching TOC logistics solutions to manufacturing employees. The fact that some workshops require computer support while others do not allows accommodating a wide range of teaching and learning styles. Your employees may be able to run these for you after study of the TOC books and tapes. Some of these games are incorporated into APICS TOC workshops, and TOC consultants and educators can provide them for you:

- Dice game from *The Goal* (no computers needed)
- Drum-Buffer-Rope (DBR) scheduling simulations conducted primarily via by personal computer and software packages (computers needed)
- APICS TOC drum-buffer-rope (DBR) games, a family of production management and scheduling simulations that are conducted in the form of "walking around and talking" exercises in conference rooms and company cafeterias and outside at picnic or beach areas (no computers needed)

## The Dice Game

The dice game described in *The Goal* helps to make the point about the need for protective capacity and time buffers if smooth operations and maximum on-time production are to be established. The game is lively, fun, requires very little equipment, and makes the critical point about the need for a "properly unbalanced plant" very nicely. Srikanth and Cavallaro's *Regaining Competitiveness* is another good source for this game.

## Computer-Supported TOC Simulations

There is an increasing number of computer-based TOC education simulators available in the industry, some available within APICS TOC workshops, some for sale directly from the APICS catalogue bookstore. Some of these simulators are provided with course materials that can be used by internal educators within companies.

## Drum-Buffer-Rope (DBR) Game Series

The drum-buffer-rope (DBR) game is a series of APICS Theory of Constraints educational workshops. The purpose of the workshops is to provide an opportunity for people in manufacturing to understand and to gain hands-on experience with the TOC drum-buffer-rope scheduling and buffer management shop floor control processes prior to introducing the changes within the actual factory environment.

## Setup for the Drum-Buffer-Rope Game

The drum-buffer-rope game should be conducted in a sizable open area, such as a company cafeteria, an outside company picnic table or beach area, an assembly room, or a conference room. The size of the room depends on the number of people participating and observing. The game can be scaled up or down as needed.

Several tables, or portions of tables, should be provided. They will be used to establish the following six simulated factory functions: purchasing and materials stores area, three fabrication cells, an assembly cell, and a finished goods stores and shipping area.

In order to staff those six functions, a minimum of six participants should be used for the workshop. If there are more than six participants, they can work with the workshop leader to divide into six roughly equal groups for the six stations. (That's equal in numbers, not necessarily in beauty or any other criterion.) If, given the sizes and shapes of the tables, too many participants are in each group, the workshop leader can divide the group into two groups: the participants in the simulation and the audience.

Prior to the beginning of the workshop, the six simulation stations will be set up as a drum-buffer-rope and buffer management operation in progress. Each of the fabrication and assembly workcenter cells will have one or more jobs in progress, a schedule of next jobs, and materials ready for some number of next tasks. The purchasing and shipping areas will have parts and required activities.

## The Action

The workshop leader will begin the simulation by walking the shipping department through one of its ongoing and routine reviews of the status of the factory's shipping time buffer (the factory's shipping-buffer). The review will be audible and visible to all workshop participants and observers.

This will lead to early identification of one or more manufactured or purchased parts that are not likely to arrive at assembly quickly enough to meet the shipping schedule. The routine and ongoing analysis of the composition of constraint and assembly time buffers will also be simulated.

These buffer management analyses will lead to simulation of a series of situations, considerations, decisions, and actions that people in the various roles make every day during actual TOC logistics operations.

New orders will be considered and sometimes added to the TOC perpetual schedule, workcenter priorities will be changed, lots will be split and overlapped, and (depending on the machine downtime and scrap experienced in the factory that day) it may become necessary to create a new TOC perpetual schedule to replace the one that has been maintained via incremental scheduling.

The format will be for all participants (and observers) to work together on each step of the thinking as the workshop leader moves the focus of the data gathering, thinking, and action from workstation to workstation. This way, all participants — and all observers — participate in all of the aspects of the operation.

The approach in this workshop is to clarify the basics, which are then done repeatedly and quickly in live manufacturing environments. In a real factory using TOC operations, these same fundamental actions would be going on for many products, workstations, and people all at once.

## Step Two: Integrate TOC With Other Management Approaches

Whenever the TOC management system is introduced, people are already thinking in terms of the other approaches to management and improvement that they have used prior to TOC.

These approaches could be Covey's *Principle-Centered Leadership*; Hayes, Wheelwright, and Clark's *Dynamic Manufacturing and Learning Organization*; Senge's *Fifth Discipline*; Deming and Juran's *Total Quality Management*; Crosby's *Quality*; Schonberger's *World Class Manufacturing*; Ohno, Goddard, Maguire, and Hay's *Just In Time (JIT)*; Toyota's *Lean Manufacturing*; Shingo's

*poke yoke;* Peters' *Excellence Management, Liberation Management,* and *Wow! Management;* Block's *Stewardship;* Gleick, Wheatley, and Malcolm's *Chaos;* Attilla's *Hunsmanship;* or *Some Combination!*[4]
     This section will provide examples of how TOC is blended smoothly with a few popular management and improvement methods and will provide some advice for dealing with all the other possible situations. The standard advice *any* TOC practitioner gives to any other experienced or new TOC practitioner for *any* important problem is, "Have at least one of the movers and shakers leading the project prepare a set of TOC logic trees that begin to identify and deal with the facts of the matter." In this case, the standard TOC advice translates to: "Have one of the movers and shakers leading the introduction of the TOC management system prepare a set of TOC logic trees that deal with the facts of the other and prior approaches to management and improvement."
     We'll come back to this. In the meantime, let's look at a few interesting examples, beginning with Dr. Covey.

## Use TOC in Support of Covey's Seven Habits

Consultant and educator Dr. Stephen Covey has popularized concepts such as *The Seven Habits of Highly Successful People*[5] and *Principle-Centered Leadership.* I admire Dr. Covey's leadership and work. His ideas strike me as having what I like to call *heart.* He comes through to me as understanding — and having — that quality of personality I like to describe with words such as *depth, dignity, character,* and *integrity.* I have never met him, but his books have the effect of what I like to call "centering" me. The calendars, audiotapes, and planning tools of his "tools for building strong families" can be found on various desks and tables in my home. So, there's no issue here of whether I think Dr. Covey's ideas are good or bad. I think they are very good.
     The question before the house, then, is over what range of situations are Dr. Covey's ideas valid and/or sufficient unto themselves, and in what circumstances does adding TOC to the mix extend the range and deliver important benefits?
     Let's begin by breaking the question down into its natural parts. In my view, Dr. Covey's ideas are

1.   Valid and useful under *all* circumstances for important, high-level purposes such as generating dignity, building character, setting tone, building morale, and building teams.

2.  Valid and sufficient for situation assessment, planning, and implementation in a wide range of very numerous and relatively simple situations.

3.  Usefully blended with TOC logic-tree processes and diagrams for situations of sufficient complexity to require a more detailed, structured, and comprehensive system to sort out the issues of situation assessment, solution formulation and testing, planning, implementation, and measurement.

Like the work of many other leadership teachers I admire, *The Seven Habits* and *Principle-Centered Leadership* systems — prior to blending with TOC — do not deliver the type of detailed, systematic, and comprehensive problem-solving solutions that come standard with the logic trees of the TOC management system. Like other leadership leaders I admire, his higher level principles just do not happen to include an effective replacement for the horrendous and industry-wide core problems of allocation-based cost accounting, an emphasis on cost reduction vs. growth, and focusing of attention on a proper expression of true economic value added.

This is an opportunity for synthesis, not a problem. As indicated below, companies who are already using Covey's methods should use TOC to add some structure to the disciplines of the seven habits as follows:

1.  *Be proactive.* This is good advice. If the person can see ahead and knows the right moves to make to be proactive, this advice alone is adequate. Since people often do know what they should be doing sooner rather than later, this will be enough to be very helpful in a great many situations. On the other hand, it is not enough if the nature of the situation is not clear to everyone (no consensus) or if the situation is so complex that nobody has yet been able to figure out what to do (or what to do that will not just be a futile gesture or a huge mistake). In the more difficult cases, just "being proactive" can amount to a spinning of wheels or even a move in a wrong direction, but with everyone still feeling better (at least for awhile) about (whew!) being in motion at all. The TOC contribution here is the use of the logic trees to work systematically through the more complicated situations. This translates the advice to "be proactive" into "stick with the logic-tree processes all the way through to the point where the prerequisites trees (PRT) are showing which intermediate objectives (IO) should be begun now, and then *do* them now!"

2.  *Begin with the end in mind.* For simple situations, individuals and teams can and should figure out — in their heads, or even intuitively — where they are headed and keep that end result in mind as they get started. For more complicated situations, TOC current and future logic trees should be used to climb systematically over the first huge hurdle of sorting out what the objective should be. TOC provides a structured approach to gaining a consensus for effective action on the nature of the problem and its most appropriate solution.

3.  *Put first things first.* Such advice sounds right and *is* right. Peter Drucker gave similar advice in his classic work, *The Effective Executive.* Others have suggested this as well.[6] This is helpful to encourage individuals and groups to fight against the natural human tendency of knowing what should be done first but doing something else first instead. Where it continues to be good cheerleading but begins to leave us empty handed on strategies and tactics is when we don't know which things should be done first. This is very important. How do we sort out which thing should be done first? The five focusing steps are useful here. The logic-tree techniques for identifying root causes and core problems help in setting priorities. The TOC throughput value added (TVA) measurement and financial systems, with associated scales of importance, help to ensure a focus on the type of growth that enables all else to be funded and to become, therefore, possible.

4.  *Think win-win.* Okay, but now what process do we use to identify the win-win solution, if not something systematic like the TOC evaporating-cloud conflict tree? There is a natural tendency to settle for compromises of the various parties' needs and wants. The TOC evaporating-cloud logic tree helps to identify breakthrough solutions which reduce the need for compromises that leave one or all parties unsatisfied with the outcome.

5.  *Seek first to understand, then to be understood.* Preparing current- and future-reality logic trees of the other party's likely view ensures that a thoughtful and complete approach has been taken toward seeking to understand. The conversations with the other party that give rise to this view communicate a human concern and sense of relationship which are important parts — beyond the logic — of what Covey is doing with this. Evaporating-cloud logic trees can be used to gain a shared understanding of mutual objectives, needs, wants, and both explicit and implicit assumptions in situations.

6. *Synergize.* How do we know that synergy has been achieved? Use future-reality logic trees that show causes giving rise to desirable effects in both parties' (or all parties') views. This demonstrates that the logic of the synergy is sound.

7. *Sharpen the saw.* Here he means continuous improvement. As discussed in Chapter 1, the opportunity is to be guided by all the TOC focusing concepts and tools — the improvement curves, three focusing questions (what to change, etc.), five focusing steps (identifying the constraint, etc.), and logic trees.

## Use TOC To Improve a Learning Organization's Foundation and Process

### Use Logic Trees To Capture Individual and Organizational Learning

I once attended a TOC case study presentation from a manager in the Mexico operations of Proctor & Gamble. He showed how use of the TOC thinking processes, specifically the logic trees, could be used to facilitate transferring know-how throughout the company. Once prepared, the TOC logic trees used for his factory's project were readily available to share with other similar factories and with group and corporate staffs with responsibilities for coordinating such efforts. A company does not have to be P&G to take advantage of this self-documenting nature of the TOC logic trees.

The APICS Constraints Management SIG, in partnership with a manager in one of the Big Three automobile companies, designed a project to characterize human resource management policies and their respective effects. This was consistent with that company's "people first!" philosophy. Data from interviews of experienced managers and executives would be captured in terms of TOC logic trees. A current-reality logic tree would set the stage for an initial condition within a factory or other group of people. Future trees would express combinations of policy factors to explore and show the likely effects in terms of the important effectiveness measurements. Prerequisites and transition logic trees would document the implementation methods and their results. This project was discussed at the 1996 APICS Constraints Management Symposium in a technical session concerning the use of TOC logic trees for field research. At this writing, due to the nature, funding, and pace of projects in the volunteer association environment, this project has not been completed. It will deliver strong benefits in terms of its nominal content as well its leading-edge demonstration of TOC methods for industry research. Parties interested in helping to build and deepen this important body of knowledge should contact a member of the APICS Constraints Management SIG leadership team.

## Use TOC To Guide Total Quality Processes

As early as January 1992, Neil Gallagher of ITT was stating that TOC was needed in order to have a successful Total Quality Management program and to implement an effective Shewhart plan-do-check-act (PDCA) cycle.[7]

At the 1995 and 1996 APICS Constraints Management symposia, U.S. Air Force Lt. General Charles Roadman and his team from the Air Force Medical Service described how they added TOC to an existing Total Quality program and derived many benefits in terms of focus and effectiveness.

Stein and Dettmer's 1997 books, both discussed earlier in this chapter, show how to use TOC principles and logic trees to provide proper priority, sequence, and coordination among concepts in quality management such as the Taguchi loss function, design of experiments (DOE), and statistical process control (SPC).

Robert Mitchell of Printronics, an experienced Total Quality manager and a Deming admirer, told the conferees at the 1997 APICS Constraints Management Symposium how he had used components of the TOC thinking processes when some of the analytical tools of the Total Quality tool set were found to provide less analytical power. Mitchell knows both the TOC and Deming methods due to his years of experience with both.

It is a useful exercise to list Dr. Deming's 14 principles and to show how the TOC principles, TOC throughput value added (TVA) measurements, and logic-tree thinking processes overlap, deepen, and extend them.

## Use TOC To Coordinate the Competing Concepts in Manufacturing Management

Most organizations face the challenge of making sense of a long list of approaches to manufacturing management. The list usually includes — just to get started — just-in-time (JIT), agile manufacturing, manufacturing resource planning II (MRP II), enterprise resource planning (ERP), world class manufacturing (WCM), dynamic manufacturing, learning organization, Toyota production system (TPS), and lean manufacturing. By now, strategies for TOC to accomplish, guide, coordinate, prevent problems from, and support all of these approaches have been worked out. I will just mention a few.

## Just-in-Time

Many of the JIT implementations outside of Japan were accomplished with the support of the book, *The Goal*. Just-in-time consultants and practitioners tell how the book helped their clients understand the relationships among the several departments of a manufacturing company, helped them to understand the problems caused by high inventories, and helped them to use inventory reduction methods with confidence.

One of the executives showing a way in which TOC and JIT cooperate is Kermit Hobbs, the manufacturing vice president for Amadas Industries.[8] One of his presentations described how his company used TOC to decide which setups to reduce. In fact, they used TOC, within a mature and successful JIT implementation, to discover that some major factory re-arrangements would make sense and actually be setup reductions.

Implementing TOC delivers all of the benefits intended from a JIT implementation, but also many more, and with fewer risks of damaging the business via taking JIT principles too far in an inflexible "purist" direction. Implementing just-in-time principles and methods, if done correctly, delivers more benefits than risks to the business. However, many JIT consultants and practitioners have taken JIT principles out of context and implemented them in ways that hurt the businesses that used them. A properly executed TOC and drum-buffer-rope implementation provides the desired reductions in inventories, lead-times, and other wastes, but also retains strategically placed inventories, deals directly with the issue of protective capacities (vs. adopting a default position of gaining short lead-times via investments in overcapacity), deals directly with current and future and strategic constraints, and provides a much stronger foundation of thinking, policy, measurement, and logistics systems.

Some JIT fans emphasize the inventory-reduction aspects of the approach, along with specific techniques such as kanbans and pull systems. Others focus on higher level principles of continuous improvement and elimination of waste, with inventory being described as one of the wastes to be eliminated. TOC's contribution to both groups is the same as to the Covey approaches: providing more comprehensive and more detailed principles, processes, measurements, and systems which allow larger and more complicated problems in a wider range of circumstances to be addressed more effectively.

## Shingo Management Methods

There have been many examples of the synergies of TOC and the Shingo improvement methods. In the late 1980s and early 1990s, the executives of United Electric Controls (UEC) in Watertown, MA, were telling of their journey to become winners of the coveted Shingo Award. Bruce Hamilton, of UEC's manufacturing organization, told an APICS dinner meeting group that his team read *The Goal* and considered it an important influence on their success.

Years later, Dave Pickels of Bal Seal Engineering presented his company's case study at the 1997 APICS Constraints Management Symposium. He told the story of how his manufacturing company had enjoyed a great deal of satisfaction and success using the methods of Shigeo Shingo, whom Pickels and many at Bal Seal admire. After a few years of increasing quality with the Shingo methods, TOC came to Bal Seal's attention. Within three months, the company had doubled profits and set the stage for continued strong growth in sales and profits.

Dr. Goldratt took industry's understanding of these synergies a step further during his keynote address at the 1997 Shingo Award conference. The Shingo Award organization is one of the primary advocates of the Shingo, lean manufacturing, and Toyota production system (TPS) methods in the U.S.[9]

## Support Your MRP II and ERP Systems with TOC's Solution to the Famous "Infinite Capacity" and "Work Order" Problems

At the 1996 APICS Constraints Management Symposium, I presented a paper entitled "The Systems Industry Is Giving Customers a Choice: Traditional and TOC Systems". This paper reviews the history of the major concepts in manufacturing software and shows how the TOC systems should be and are replacing the closed-loop MRP (also known as little MRP, mrp 1, and materials requirements planning) and CRP systems, while retaining and fitting within the MRP II and ERP architectures and processes. The paper includes a chart with a side-by-side analysis of why TOC finite-capacity systems are replacing the infinite-capacity MRP/CRP systems and shows how the TOC and MRP II/ERP systems work well together.

## Use TOC To Guide and Shape Meaningful Coincidence

Every high-performer knows that the act of making a decision and the attitudes of determination and persistence cause one to notice the arrival of information and resources that seem coincidental, except that it always happens that way. After awhile, the coincidences seem to take on a pattern, seem to have some meaning. James Redfield is among the many who have noticed the phenomenon and has written a few books about it, including the bestseller, *The Celestine Prophecy.* One of his central themes — nevermind how he gets there, by the way — is the nice phrase "meaningful coincidence".

There are a lot of quotations in the culture that have similar meanings, such as "knock and it shall be opened unto you" or "ask and ye shall receive". There is another one attributed to Goethe along the lines of "Boldness has genius, power ... a whole stream of ... coincidence."

The point is that meaningful coincidence might not be everything, but it *is* something. Let's use it over the range of its validity. We do not need to take this to some awkward level simply to do what works.

We can benefit from using TOC principles and logic trees to fine-tune our definition and perception of the objectives to which we are applying our determination, will, and faith. In so doing, we can at least indirectly influence the stream of incoming meaningful coincidence.

## Some General Advice: Other Management and Improvement Approaches

Virtually every other management and improvement approach can benefit from the addition of TOC. The challenge is to find the way to communicate that to the experts and promoters of other methods within your organization.

There are so many approaches to management and improvement that we cannot possibly cover all of them here. I have covered a few in this section to give an idea of how to do it. Chapters 1 to 4, the APICS Constraints Management symposia audiotapes, and many of the footnotes to this book provide additional ideas for integrating TOC smoothly into environments that have been using other methods.

The ideal outcome is that, during the initial education work (Step One), the experts in other methods see the opportunity for themselves to integrate TOC principles and tools into their approaches. Since that will not necessarily always happen, this integration of TOC with other management approaches has been established as an activity of its own (Step Two). If you get the needed outcome easily, without planning and implementing a Step Two, fine. Enjoy it and move on to Step Three. If you have to work at it, then you have to work at it. As mentioned in the beginning of this section, somebody in the TOC leadership group should prepare a set of logic trees that deal with the facts of the prior management methods in *your* organization. Good luck!

# Step Three: Build a TVA Financial Management System

Do this in two steps:

1. Build consensus as to what is going to be done.
2. Build the TOC throughput value added (TVA) accounting system.

To build the consensus, you may need to...

## *Invite Detective Columbo[10] in To Introduce TOC Accounting*

This part is important, so be sure to give this job to somebody who can carry it off properly. You need to assign someone to dress up in rumpled shirt, tie, trousers, and overcoat. He or she needs to have tousled hair, posture that is bent over slightly, and a not-quite-scatterbrain way of squinting eyes, pursing lips, and waving the hands around to make a point. If you can get the real Peter Falk to come to your management meeting, that is obviously best. If not, work with your best substitute. Here is what Detective Columbo needs to do for you.

[Columbo begins by sticking his or her head into the conference or assembly room door and asks something like ...]

*Columbo:* "Excuse me. Excuse me, ladies and gentlemen. I'm very sorry to interrupt, but is this the investigation into the company's cost accounting?"

[It's best if the leader of the meeting has, within the last half hour, turned the subject to cost accounting. In the confusion, everybody will be looking at each other. Nobody will give a straight answer to his question. Columbo should indicate that he has suddenly realized what he has done, bounce the fingers of his right hand off his forehead, and begin sheepishly and profusely to apologize — but he should never leave the doorway or room. The talk should go something like ...]

*Columbo:* "I guess not. This must be the wrong room. You know, I'm very sorry, ladies and gentlemen, very sorry that I interrupted your meeting. You were probably talking about something very important. Yes, I can see you were discussing something *very* important, and what did I do? Like a dummy, I stuck my head in the door and made a big distraction. I distracted you from talking about ... [to the meeting leader] what *were* you talking about, sir [or ma'am, as the case may be]?

[The leader, who knows it is a scam, makes an embarrassed face, as if not sure how to get rid of this guy who's suddenly in the middle of the meeting.]

*Columbo:* Really, what were you talking about, sir? [Pause] I hit the nail on the head, didn't I, sir? You were talking about cost accounting, weren't you?

*Leader* [uncomfortable]: Yes, we were talking about cost accounting, but ...

*Columbo* [interrupts]: Well, sir, I have to tell you, that's great news. You know why? I'll tell you why. Because I'm having a very hard time understanding something about cost accounting, and I'll bet you and your friends here can help me understand where I've been going wrong in my thinking.

*Leader* [feigning exasperation]: I'm really very sorry, Mr., Mr. ...

*Columbo:* Columbo, sir. Detective Columbo, at your service.

*Leader* [stunned]: Detective Columbo? Here?

*Columbo:* Yes, sir. Will you help me with this problem about allocating overheads and calculating product costs, sir? It's really baffling me. I'd really appreciate it.

*Leader* [suddenly very impressed and flattered]: The real Detective Columbo? [Then, remembering he needs to answer the question, ...] Why, certainly. I'd ... we'd ... I'm sure we would all be very glad to help you, Detective, [to group] wouldn't we? [Group agrees to help.]

*Columbo:* Thank *you* very much, sir. You don't know much help this could be to me, to have all you experts in cost accounting help me get these things straight in my head. I was working with a company the other day that made ... What *did* they make? You know, sir, I forget what they made. Well, nevermind what they made. Do you people make things? [Somebody indicates yes.] You do? That's great. So you must have a factory. Am I right? [Yes, again.] What do you make?

*Leader:* We make pumps.

*Columbo:* Pumps. That's very interesting, sir. I have a pump in my house. Do you believe that, sir? [Leader indicates he believes it.] Well, it's true. I really do. And do you know what happens sometimes at my house? Sometimes, when it rains, the water leaks up from under the basement floor and makes a big lake in my cellar. [To one of the people in the meeting ...] Did that ever happen to you, ma'am? ["Ma'am" indicates no.] Well, you're a lucky lady, ma'am, because the only way I can get that water out is to use my pump. That's right, I pump the water out the window. Is that the kind of pumps you make, ladies and gentlemen, the kind people use in their basements? No? Okay. What kind of pumps *do* you make? Okay. Well, nevermind my kind of pump, let's talk about the ones you make, because, to tell you the truth, it's not the pumps that confuse me. I understand the pumps. It's the overhead allocations in cost accounting that confuse me.

*Leader* [pretending to become impatient at the wasted time]: How do you think we can help you, Detective?

*Columbo:* Let me ask you a few questions. You know how much money comes into the company for each pump that's sold, don't you?

*Group:* Yes, we know the sales price. If there are any discounts, we know them, too, and can subtract them to calculate the net sales price.

*Columbo:* I see. And you also know how many pumps were sold at which net sales prices in any given day, week, month, quarter, or year?

*Group:* Absolutely.

*Columbo:* So you know how much money came into the company from sales of pumps in any time period?

*Group:* Yes. And if we're talking about the future, a scenario we're exploring, we know the assumed number of each model sold and the projected prices at which they will be sold. But what are you driving at?

*Columbo:* Be patient with me, ladies and gentlemen. Like I said, I'm pretty confused about all of this complicated cost accounting business. I figure, if I can keep it simple, I might be able to understand. I really hope you can help me do that.

*Group:* We'll try.

*Columbo:* Thank you very much. You don't know how much I appreciate this. In fact, I was telling Mrs. Columbo just the other day that I hoped I'd run into a nice group of experts who really understand this cost accounting so I could get it all explained to me, because I must be missing something. I'll show you what I mean. You make these pumps out of parts, right?

*Group:* Right. We buy raw materials that the factory turns into finished component parts. We also buy purchased parts like screws and O-rings that are used directly in assembling the finished pump.

*Columbo:* That sounds just like what I found at the other factory I visited the other day. So that means you know how much money went out of the company to buy raw materials and purchased parts for each unit, is that correct?

*Group:* Yes. That money goes from our company to the supplier companies.

*Columbo:* Is there other money that goes out from the company for each pump that's made?

*Group:* Yes, every pump has parts that need to be heat treated, coated, or chemically cleaned by outside service vendors. Plus, there are certain packaging, shipping, and commission costs that we pay for each pump we make and ship.

*Columbo:* Those would be the totally variable costs (TVC). That's very interesting, very interesting indeed. You know how much money is coming into the company from sales and how much money is going out of the company for TVC for each pump you make.

*Group:* Right.

*Columbo:* The TOC people say subtract TVC per unit from sales per unit to get TOC throughput value added (TVA) per unit. Let me write that down right here on this flipchart. Is it okay if I use this flipchart, sir? Thank you. Thank you very much. Here it is

$$\text{Sales}_{unit} - \text{TVC}_{unit} = \text{TVA}_{unit}$$

We could do that for all of the pumps added together, is that correct?

*Group:* Correct again, Inspector. What about the overhead? We need to allocate the overhead to individual products, and calculate allocation-based product costs, in order to know the full economic picture.

*Columbo:* Interesting that you say that. That's what the people at that other company said, too. And that's what's been bothering me. Stay with me a minute. Because I'm going to get to a picture that tells the whole story, and I get there before I ever allocate a single bit of overhead to a product cost. That really bothers me that I see a complete picture before I allocate anything. Can you see why that bothers me? It bothers me because it makes me wonder why people are allocating to get an answer they already have before they allocate anything. Do you see my problem, sir? Ma'am? So before we get into this labor and overhead thing, maybe you can tell me something. About how much money, TVA, or sales-less-TVC flows into the company for each pump?

*Group:* We sell each of these pumps for $500, net of discounts.

*Columbo:* Okay, that's sales. How much money goes out for a pump?

*Group:* Some of the parts are manufactured from raw materials purchased from suppliers and then assembled into the final pump. These raw materials cost $100. The purchased parts we use directly in assemblies cost another $50.

The other totally variable costs amount to another $50. The totally variable costs, then, that we pay to outside parties for each pump amount to $200.

*Columbo:* So that means that once you've collected the money from the customer, and paid the suppliers, the company has $300 more dollars of TVA available to spend on things than it had before you made and sold the pump, is that correct?

*Group:* That's correct.

*Columbo:* That TVA is good stuff too. You could say that TVA is a lot like the cash flow from operations that makes continuing operations possible. That's a pretty important thing to pay attention to. Now I can multiply that $300 by the number of pumps we expect to ship in, say, a 5-year planning period?

*Group:* Well, yes and no. The $300 only works for the pumps sold in the Fire Truck application market. The TVA per unit is different for different models, for different applications, in different geographic areas. It can also be different in different years because we are constantly updating and improving our products and processes, working on reducing expenditures for raw materials and purchased parts and outside services, and increasing the amount of product we can make with our factory.

*Columbo:* That makes sense to me. May I use another page on your flipchart, sir. Thank you very much, very much. Let me draw a chart here. Stay with me. I'm not very good at this. But let's say we make a chart of your product lines in the first column on the left. Then we'll show the different geographic or application markets, whichever produce important differences in unit prices or totally variable costs. Then we'll show the results of the plans and projections for pricing, manufacturing capacity, and raw materials expenditures. For purposes of this discussion, let's make assumptions for how many pumps your company will sell into each of those market areas in each year. That will give us a 5-year strategic planning chart that shows the total TVA for each product line in each market niche that looks like this (Figure 6.1). Each box in the TVA section of the chart is calculated as:

$$\text{TVA}_{\text{per unit for a product in a market niche}} \times \# \text{ of units sold}$$
$$= \text{TVA}_{\text{total}} \text{ for that product/market/period (year)}$$

[This gives rise to a larger version of Figure 6.1, which has been made up ahead of time so that it can be pulled out after Detective Columbo has made it clear how it is to be prepared.]

## The TVA Financial Management System for Strategic, Product, and Improvements Planning and Control

| Product Model | Market (Application) | Market (Geographical) | TVA Year 1 | TVA Year 2 | TVA Year 3 | TVA Year 4 | TVA Year 5 |
|---|---|---|---|---|---|---|---|
| X-Pump | Ship | North America | 200 | 210 | 220 | 220 | 230 |
| | | Europe | 200 | 200 | 200 | 200 | 200 |
| | | Asia | 200 | 200 | 200 | 200 | 200 |
| | Truck | North America | 500 | 500 | 500 | 500 | 500 |
| | | Europe | 400 | 410 | 430 | 440 | 450 |
| | | Asia | 400 | 400 | 400 | 400 | 400 |
| | Fire | North America | 300 | 300 | 300 | 300 | 300 |
| | | Europe | 300 | 310 | 320 | 330 | 350 |
| | | Asia | 300 | 300 | 300 | 300 | 300 |
| Y-Pump | Ship | North America | 200 | 250 | 350 | 400 | 450 |
| Total TOC value added | | | 3000 | 3080 | 3220 | 3290 | 3380 |
| Operating expense (OE) | | | 2000 | 2000 | 2000 | 2000 | 2000 |
| Net profit (NP) | | | 1000 | 1080 | 1220 | 1290 | 1380 |
| Investment | | | 5000 | 5000 | 5000 | 5000 | 5000 |
| Capital charge (15%) | | | 750 | 750 | 750 | 750 | 750 |
| Economic value added | | | 250 | 330 | 470 | 540 | 630 |
| Required ROI | | | 15.0% | 15.0% | 15.0% | 15.0% | 15.0% |
| Additional ROI | | | 5.0% | 6.6% | 9.4% | 10.8% | 12.6% |

*Note:* All figures are 1000s. TVA for a product market segment for a given year is calculated as sales per unit less totally variable costs per unit times number of units sold in the year.

**Figure 6.1. Detective Columbo's TVA-I-OE Chart (TVA-I-OE Accounting, a Portion of the TVA Financial Management System for Strategic, Product, and Improvements Planning and Control)**

*Group:* Good so far. Now, we need to take the labor and overhead out of the line called "Operating expense" in order to calculate more accurate product costs ...

*Columbo* [interrupts]: Right you are, Ma'am. I almost did it again. I almost forgot about that labor and overhead part again. You know, I think the reason I keep forgetting to allocate overhead is because that's the part that's really been bothering me.

*Leader:* Why does that bother you, Detective?

*Columbo:* I'll tell you why that bothers me. Look at that chart right there. There's the picture of one potential business plan, right there in front of us. The company knows the TVA for each product in each market. It must know or be able to project production and office salaries and benefits, advertising expenditures, and other expenditures that don't vary directly with units. Otherwise, how would it get numbers in the operating expense line?

*Group:* True. For a given business scenario, we have to estimate how many people and what expenditures are required in all areas of the company and how our reengineering and other improvement activities are going to affect that. That shows up in the operating expense line. So we can now take parts of that operating expense and allocate it to products to form allocation-based product costs. Some of the overhead can be argued as varying somewhat with units of product in the mix. Most of it can't. But where there is no relationship, we always just make assumptions and stick with them.

*Columbo:* Even though it doesn't make sense? In other words, even though the allocations don't reflect the true underlying economics.

*Group:* Right. But we have to allocate the costs to some product.

*Columbo:* Why?

*Group:* Because we always do.

*Columbo:* But why? You have the entire economic picture in front of you here. You know the cause-and-effect relationships between what you plan to do and the levels of operating expense in each year, in excruciating detail, at the right level of precision. If you change the assumptions in the mix of product, you can also change the operating expense levels, if there are such effects. If you had allocated the overheads to products, you would have to have unallocate them, to then re-allocate them to reflect the new mix, right?

*Group* [thinking about this]: Hmmm.

*Columbo:* Why do you do that to yourselves? Sure, change the assumed mix, change the TVA figures calculated from the assumed pricing and true variable costs, and change the operating expenses to show the new full economic picture. But why then add the work of allocating some of those operating expenses to create allocation-based product costs? You get no additional value, you do more work, you mask the true economics, and then — at the end of each period of operations — you have to deal with variance analysis, including unraveling "mix variance" in order to close the books and make reports, right?

*Group* [still thinking]: Right. But there's another issue. There's not just the issue of the information we should use for making decisions. There's also the allocations of labor and overhead we have to make for inventory accounting, for purposes of tax reporting, and for calculating the cost of goods sold for the financial accounting that conforms to generally accepted accounting principles (GAAP).

*Columbo:* Absolutely. For external reporting, you need some allocations. I wonder if the authorities realize how much extra effort they're creating for no useful purpose and that the allocations to inventory and cost of goods sold allow managers to manipulate reported profits in ways that hide rather than report true underlying economic activity. Do you think that might change someday, Ma'am? You think so? That's interesting, Ma'am. In the meantime, internal management accounting is not subject to those same external rules. You are not required to go to more work and effort in order to create less clarity and greater potential for mistakes, in order to go to even more work at the other end of the accounting cycle, or for mix change assumptions in planning cycles. Isn't that true?

*Group:* That's right. There's no reason we can't use a spreadsheet or a database reporting format to give ourselves a clear TVA view for use in strategic decision-making, deciding which improvements to focus on in the company, and projecting the economic effects of our plans for investment, marketing, product development, and continuous improvement.

*Columbo:* So that's what's been bothering me, ladies and gentlemen. We can know the true life-cycle economics of a new product, product family, factory, any other change, and of the entire business — under any circumstance we can imagine — and still never have allocated a single bit of labor and overhead to form a single fictional math entity called an "allocation-based product cost". Yet, companies continue to do it! Why is that, ladies and gentlemen? That's what's been bothering me. Why is that?

*Group:* Okay, okay. We'll stop creating allocation-based product costs for making decisions and start to use the TVA financial management system.

*Columbo:* You know, Ma'am, it's a funny thing about what you just said. That's the same thing that last company said after I asked them the same bunch of questions about all this complicated cost accounting business. In fact, it's the same thing every company says when I ask them these questions. I wonder why that always happens. Well, anyway, I really want to thank you for helping me think about this overhead allocation business. Mrs. Columbo is going to very happy to hear about this. Goodbye. [Columbo exits the room].

After Detective Columbo leaves, there may still be a few questions. For now, let's deal with just one of them.

## Why Does TOC Not Treat Factory Direct Labor as a Variable Cost?

Several reasons. The first is that it simplifies and clarifies the management accounting reports. The firm gets a more clear picture of the company-wide economic impacts of decisions, while not spending a lot of time debating over and applying allocation bases and rates and then reconciling various types of variance. More importantly, in some circumstances such as in union environments, factory labor is not variable at all and behaves more like period operating expense.

Of primary importance, however, is that companies begin to treat factory labor as not variable at all — even if no union contract or any other legal obligation requires it. Treating factory labor as a resource that is not going away as unit production and sales go up and down (the equivalent of treating the costs of factory labor as period operating expense vs. variable labor cost) is a key element of creating determination within a company to keep a constant focus on maintaining and increasing TVA. This is consistent with the TOC strategic planning process and the TVA world scale-of-importance policy to maintain a constant priority on growth of the TVA flow of funds into the business.

This approach and this TVA management accounting system focus efforts and increase the likelihood that employment stability will be high. If layoffs have to happen, due to the failure to prioritize growth efforts or to make the efforts effective, then the adjustment to account for the layoffs is made by adjusting period operating expense.

## Next Step: Build the TVA Financial System

In time, all the providers of accounting and decision-support software will provide direct and fully integrated support for the TVA financial management system. Until then, do as the all the pioneering TOC companies have done and build your own with internal data systems resources.

I have built TVA reports — for the executives of manufacturing companies who had responsibility for several hundred million dollars of sales — on early versions of the Lotus 1-2-3 spreadsheet. It is even easier today with the increased power offered by the newer Lotus products, Microsoft Excel, Microsoft Access, Microsoft Visual Basic, all the so-called fourth generation language (4GL) and high-level application building software, and all the relational database technologies and report writers. People can even resort to (heaven forbid!) sequential program code and any other approach to semi-custom or full-custom solutions.

If the TVA system can be simple, even on the back of envelope, then make it simple. If it needs to be more comprehensive, in order to match the opportunities of the business, then make it both simple and comprehensive.

Companies at the high end of scale and complexity may need to call on the TOC consulting and services lines of business of the major accounting firms or systems integrators in order to get the right TVA financial management system properly integrated with the TOC supply chain, scheduling, project management, and logic-tree networked learning organization tools.

## *Don't Make Management Accounting Difficult: Use TVA*

There it is: the heart of the TOC TVA financial management system. If you always thought management accounting should be that simple, you were right.

# Step Four: Use TOC for Projects Other than Major Changes to the Logistics Infrastructure

## *Keep Track of TOC Interest and Use*

During the TOC education, TOC integration, and TVA measurements stages (Steps One to Three), keep track of TOC interest and spontaneous uses. When the time is right, call another management meeting to create one or more projects to deepen the team's skill in use of TOC logic trees.

This is an important point for a company that is serious about introducing TOC as the foundation for its education infrastructure. This step is important for many reasons, but two stand out:

1.   It ensures that executives understand the simplicity, power, and chal-lenges involved in the use of the TOC logic trees.
2.   It avoids punishing people for doing the right thing.

## *Encourage the Executive Management Team To Start Using the TOC Logic Trees*

This will ensure that executives understand the simplicity, power, and chal-lenges involved in the use of the TOC logic-tree thinking processes for specific projects. All the executives will know, from first-hand experience, that the TOC logic trees increase precision in thinking. They will know that anyone can use them, to one extent or another. They will recognize the incredible value that the logic-tree thinking processes provide in increasing the effectiveness of meetings. They will be able to serve as role models by showing their own partial trees or summary and communication trees prepared at high levels of aggregation.

They will be able to help in making the standard, but important, judgments about what is written down vs. what is not, about how things are stated for which audiences and when, and about what levels of detail are most appropriate to effectiveness. These are among the most important benefits that the skills and experience of executives deliver to organizations. These are as needed in this new approach to analysis, planning, implementation, and communication as they are in any other.

They will be able to use and oversee the use of large TOC logic trees for coordinating large projects. They will be able to understand the need for TOC education to increase "knowledge worker" productivity throughout the company. They will understand the tools, traps, and solutions in the several major logic-tree thinking process flows.

On the one hand, there are things that can be learned simply as improved technique or as good advice. The *Mission Impossible* example in Chapter 1 happens to contain a very great deal of this type of obstacle pre-clearing advice that has not been worked out before or been written anywhere else. One or a few TOC experts may disagree with that advice, especially the part about not getting tied up in your shoestrings by, in turn, getting hung up too early in the detailed formal rules of logic trees. They will call this approach an "error" rather than "Tom's interpretation" and they will be wrong.

On the other hand, there is simply no substitute for just diving in and working with the logic trees. As with any other tool and method in life, each person has to work our his or her own style of working with logic trees. Some people work at them intensely for sustained periods and all the time. These people will be good at handling one type of situation and not as good at handling others. Other people will build rough overall trees quickly and then move on to another activity for a while. These people will make frequent or infrequent small additions to their library of logic trees every once in a while, having taken note of ideas that flashed into their minds when they were working on other projects in other places. Some may have support staff do the actual drawing, typing, and inevitable re-drawing and re-typing of the individual or group logic trees.

As with all other tools and methods in life, executives will understand better how to make judgments on this "different strokes for different folks" issue if they have tried, failed, tried again, succeeded, tried again, failed again on larger problems, tried again, succeeded on larger problems, and so forth — just like the rest of the employee population will be doing.

They will better understand that the TOC tools increase the effectiveness of people who are already committed to living out the notion that one person can make a constructive difference. The same tools are excellent vehicles for individuals interested in cultivating a "can-do" attitude by developing the associated skills.

## Every Manager Develops Trees

Each member of the management team should prepare his or her own initial current-reality, evaporating-cloud, future-reality, prerequisites, and transitions logic trees. Some individuals will be less comfortable, some more comfortable, with the logic-tree methods. This is normal. Everyone should do at least one of each in order to understand the nature, value, pros, cons, and issues involved in their use. Just as not all people in a company like or are proficient in all of the same things, people will differ in their preference and aptitude for the TOC logic trees. No problem. This is not intended to become a situation like the one created by the famed Procrustes, the guy with the one-bed-fits-all problem.[11]

As the trees are used, the company will benefit. For example, lots of people do not like accounting or expense reports, but they use them and know they are being used around them, and the company and even the complaining people get the benefits from their use. As another analogy, the brilliant accountants play important contributing roles. The brilliant logic-tree meisters will do so, too. You only need one or a few really strong logic-tree builders in an organization, along with a whole lot of people who can handle them reasonably well, along with everybody else in the organization who at least understand the model of thinking and can speak the language. So it is with every other role in life.

## Role Models for Executive Use of TOC Thinking Processes

There are three excellent examples of this in the public domain and discussed in this book:

- Kent Moore Cabinets, a small company described, among other places, in the Institute of Management Accountants' TOC field study report

- Harris Semiconductor, a large technology corporation whose cases appeared in the Bob Murphy and Sharon Sines portion of the proceedings for the 1996 APICS Constraints Management Symposium and were discussed in Goldratt's 1997 APICS Constraints Management Symposium presentation about TOC project management
- U.S. Air Force Medical Service, a military and service organization whose case was presented at both the 1995 and 1996 APICS Constraints Management symposia

## Early Executive Use of TOC Logic Trees To Maintain and Increase TVA Avoids Punishing People for Doing the Right Things

By acquainting the entire top management team with TOC at the beginning of the serious implementation process, a company reduces its risk of sending destructive messages by making the oft-repeated mistake of punishing people who have done the right thing for the company. Here is what has happened so many times:

- Somebody in the company reads *The Goal*, likes it, and persuades others to cooperate in putting the ideas to work.
- The ideas work.
- The group that did the work, usually the production portion of the operations organization, now does more with less.
- The rest of the company thinks using TOC is a "production thing" and has not been preparing itself to increase sales in response to the work in production.
- The business cycle softens.
- The company comes under pressure for profits.
- Since profits are not going up due to increased sales and TVA, the company seeks the profitability wherever they can find it.
- The group with the "excess" staffing is the one that used the TOC to improve. (This is not excess capacity. Part of it is current protective capacity, which, when trimmed, will upset the delivery performance and orderliness of the operation, which will make profitability problems worse, which will increase pressure for more staffing cuts. Most of the rest is the productive capacity and protective capacity that should have been used for what should have been the next increase in sales and TVA cash flows). The production department people

who did improve via TOC to create lean and effective operations — ready to support additional growth in sales, TVA, and profits — experience the layoffs. Meanwhile, the sales, marketing, engineering, and other divisions that did not proactively improve are still so busy — at not being effective in getting the company's existing and new products sold in greater volumes into additional markets at attractive prices — that they cannot possibly be spared. They are not touched by the layoffs.

- The message is unintentional, but clear: Don't cooperate with improvement efforts, TOC or otherwise.
- The company is now worse off than it was before.

There was a period in the early 1990s when Eli Goldratt saw this happening in large numbers. He instructed his network of partners, franchised associates, and licensees to refuse to support requests from industry for support in TOC implementations if conditions were not right for maintaining or growing TVA to prevent this problem.

## Use TOC Logic Trees and TVA-I-OE Measures for Strategy

Applications of TOC principles, thinking processes, and information systems in support of strategic planning have been covered in my papers presented at the 1997 APICS Constraints Management Symposium and the 1997 Flower City business conference. Companion reads include Peter Schwarz's *The Art of the Long View* and Michael Porter's many fine works on strategy.

## Use TOC To Improve Competitive Monitoring

This is related to strategy and becomes a project of its own. Many important situations require the ability to collect and evaluate information concerning dynamic and complex environments. Details that may seem insignificant to the layman must be noticed and interpreted by the experts in order to gain timely notice of major decisions, events, or changes.

### Commercial, Legal, Diplomatic, and Military Applications

Industry analysis and monitoring is one important application. Competitive intelligence gathering — in all its commercial, legal, diplomatic, and military contexts — is a related application.

The questions of which observations are to be made and which interpretations are to be made of the resulting data may be answered by a single individual (in the case of a small business) or by a large team of experts (in the case of industrial, large-scale litigation, diplomatic, or military situations).

## More Effective Use of Brilliant Strategists

The expertise of a single expert in an area may be extended and made much more effective through the use of TOC current, future, and transition trees. Here's how.

## Mutual Fund Industry

Let's use an example from a commercial context. If you work in a law firm or government agency, you can make the translation readily from this example. Let's use an industry with which most readers will have at least a passing acquaintance — the mutual fund industry. Many readers will own stocks, bonds, or other mutual funds, either directly or via company 401(k) or other tax-deferred savings plans. Mutual funds can be purchased as "load" or "no-load" funds. It is basically a question of getting what you pay for, in that the investor is charged a fee when buying load funds in return for the services provided by the broker selling the fund shares. In the case of no-load funds, the investor determines he or she does not need much advice or assistance from the broker and pays no fee to the broker at all. So mutual fund brokers earn their incomes, in part, by providing services that justify their being paid the fees associated with the load mutual funds. There are quite a few brokerage companies, including familiar names such as Merrill-Lynch, Dean Witter, T. Rowe Price, and many more.

## The Scenario

On a morning in June, Joe, a broker at the BuysEmLow mutual fund brokerage company, opens his copy of the *Wall Street Journal*. He reads that one of his company's major competitors, Smith-Barney, a major brokerage, has announced it will now begin to offer no-load mutual funds in addition to its load funds and other investment options. He takes the paper over to his boss and asks, "Are these people crazy? Don't they make their money by selling loaded mutual funds? Aren't they putting themselves out of business by making a move like this? And, by the way, what's the impact of this going to be on us and on other brokerage companies who rely on selling load mutual funds?"

His boss listens carefully to the question. She considers for a moment and begins to share her analysis of the ramifications for investors, for Smith-Barney, and for other brokerage firms. However, as she begins to speak, she decides this may be of sufficient importance to their industry that a casual and verbal analysis ultimately will not suffice. Because a clear and correct view will be needed eventually, she determines to kill two birds with one stone by capturing the thinking that emerges in their discussion in terms of cause-and-effect linkages. Figure 6.2 is the result of the quick brainstorming session. They have prepared these logic-tree diagrams before. They use them to ensure that investment combinations meet the objectives and are in the best interest of certain clients. They are then useful for communicating the wisdom of the selections.

## Use TOC Best Practices in Strategy

Use the TOC best practices in strategy. For example, treat "money" (current and future TVA-I-OE, net profit, and return-on-investment) as the operational goal and employees' and customers' needs as necessary conditions of equivalent importance as the goal. See *It's Not Luck* for examples of the TOC best practices in strategy and of the logic processes which support them.[12] This topic deserves book-length treatment and will get it, but the following are among the TOC concepts that are used for strategic planning:

- Segment the markets, not the resources, to prevent economic downturns in any one market from creating pressure for layoffs.
- Position products in several markets for which prices do not affect each other to increase likelihood of steady sales, steady TVA, and the ability to price well.
- Draw a current-reality tree of the market.
- Determine the core problem of the market.
- Draw communication trees.
- Draw transition trees for sales processes.
- Realize that layoffs are more expensive than they seem, even to the firm, but most certainly to society.
- Position product offerings vis-à-vis price sensitivity.
- Analyze customer selection.

## TOC Current Tree
### Assessment of New Event in Mutual Fund Industry
### (Smith-Barney begins selling no-load mutual funds)

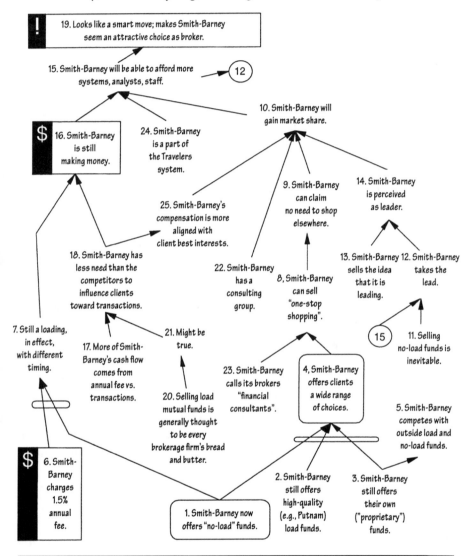

**Figure 6.2 TOC Current-Reality Tree (CRT)**

Use TOC principles, logic trees, and information systems to assess and improve some aspect of the company's strategic positioning beginning with current-reality tree processes. Sometimes strategic uses of TOC will arise because of some problem. By working with TOC logic trees on a demanding symptom, the solution that is implemented can be combined with other and more fundamental cures to gain strategic impact.

Sometimes the best place to start is with some particular new idea. Treat this as a vision statement and use TOC future-tree evaluation processes to test the vision. Again, although the idea that triggers the thinking may not be strategic, dealing with the idea using logic trees can give rise to progress on more fundamental matters of strategic significance.

## Use TOC To Improve a Process

### Use TOC and TP as the Process To Select Processes for Process Management, Improvement, or Reengineering

Be careful not to let the process management, process improvement, and process reengineering communities kill off your TOC momentum. There is a narrow focus that populates some corporate finance departments that takes an overly simplistic view because they can control it, and because it maintains their hold on the organization.

Not all financial managers are like this — not even most — but the ones to beware of are the same ones who killed off your company's growth by focusing everybody's attention on inventory and operating expense reduction at the expense of growth efforts. They are the same ones who will always prefer the riskless process they can dominate to the still safe, but undominated, open-ended process that can lead to an improved bottom-line and organizational health.

This is a situation where, unless they know what they are looking for, top management can be persuaded that "more results" are being obtained with the process flow mapping and activity costing activities. The reality is that any approach other than beginning with the scope of the overall system and working from that perspective can and will cause efforts to be either too narrow or of only local impact. The more narrow efforts consume a lot of time with pretty project plans, show that easy schedules have been met, and generate lots of paperwork with lots of numbers that look precise but are so laden with assumptions that they miss the point.

It *is* true that processes should be improved. It is not true that it is easy to separate one process from another when the purpose is to have a bottom-line impact on the organization. It is *not* true that focusing on a single process is effective when the bottom-line results needed should really be sought from a collection of changes from several areas.

The way to handle this is to consider TOC as the company's process for focusing ongoing improvement. As discussed in Chapter 1, TOC's focusing process and logic trees become the company's process for selecting processes and parts of processes to improve.

## Effective Business Process Design

In order for the company to be supported in accomplishing correct actions to serve its targeted customers, business processes must be effective (entity 19 on Figure 1.1). These processes can range from manufacturing process design decisions — e.g., cells, automation, setup procedures — to business planning, sales and operations planning, demand management, master planning, scheduling, and shop control. If TOC thinking processes (the logic trees) are in regular use (entity 7a), any need for changes to business processes will be anticipated. When these thinking processes are in use in combination with the determination to work on high-leverage tasks (entity 6), the risk of frittering away time and focus on too many projects is greatly diminished. The result is that, when the changes to business processes become necessary to support changes in the role and focus of factory capabilities, the organization works on deep causes instead of symptoms and gains the benefits of sound business processes.

## Use TOC Thinking Processes in Place of Benchmarking and Surveys

Enormous amounts of time and money can be consumed in formal benchmarking and survey activities. The fundamental value of the entire benchmarking exercise most often can be gained with a fraction of the people, the cost, and the lost time. Use the TOC logic-tree processes to accomplish this by using cause-and-effect thinking to arrive at the best practice for your company, no matter what all the other players in the industry are doing. This does not mean a company should keep its head in the sand. It is okay to peek at what others in the industry are doing by working the grapevine, reading business publications, and using other methods of industry and competitor research. This can be useful in the work with the logic trees. But that is a lot different from organizing teams to be sent to other companies and incurring the obligation to have other companies visit your own operations, with all the associated time spent to manage the processes and other risks.

### Encourage Use of TP in Management Skills Format

Due to the well-known applications of TOC to marketing, measurement, project management, logistics, and scheduling, many people miss the inherent value that TOC provides in the area of basic management skills: empowerment. Study the logic trees and use them in day-to-day managing.

## Step Five: The Major TOC Logistics Projects

### Build Drum-Buffer-Rope and Other Useful Logistics Systems on Your Company's New Foundation of TOC Processes

If you have followed the first four steps, you are ready for this one. There is a great deal of literature and education available to assist you in this area, as described earlier in this chapter and in Chapters 3 and 4. More is on the way.

### Establish TOC Information Systems To Support the New Logistics Infrastructures

As in the case with the accounting solutions, some companies may be able to get a great deal of progress from the use of a few simple spreadsheet or database programs. Other companies may need to acquire the TOC services division of major accounting firms or systems integrators to build the TOC infrastructure of networked logic-tree thinking process diagramming tools, project management, TVA financial management system, and TOC production and supply chain systems.

Whatever works. Once again, as discussed in this book, there is a great deal of additional information and education available (with more on the way) to assist you with this part of your project.

## When the Five Implementation Steps Are Complete

When the implementation sequence has been accomplished, the company goes about its business. More often than not, and more often than most companies, the company pleases its shareholders, satisfies its customers, grows enough in TVA to fund continued growth and reasonable employment stability, and maintains a work environment in which the use of common sense is perceived by the loyal and thoughtful subset of employees. In other words, the company enjoys the benefits of having wisely invested in that new and practical expression of management science, the Theory of Constraints, and in the emerging global standard approach to management known as the TOC management system. Congratulations!

So that is where we are headed. How did we get there? The answer, of course, is the five implementation steps covered in this chapter, namely:

1. Conduct TOC education.
2. Integrate TOC with other management approaches.
3. Build a TVA financial management system.
4. Use TOC principles and logic trees for projects outside logistics.
5. Build the TOC logistics systems infrastructure.

## What If My Company Is Not Introducing TOC?

That leaves us with another problem. Those five steps assume that the entire company is introducing TOC as a management system. What can people do in the meantime, between (1) learning about TOC and (2) finding themselves in an organization that uses TOC principles and processes as standard day-to-day tools? What will increase understanding and skill with TOC, especially the principles and logic trees, and also increase the likelihood that the company will introduce TOC more quickly? The answer comes in seven parts, discussed in the next section.

# Seven Things You Can Do Right Away!

Many aspects of introducing the TOC system will take time, but there is no reason to delay doing the following list of seven things. If you are one of the people working to persuade your company to introduce the TOC management system correctly, here are things you can and should do right away.

## Do These Seven Things Right Away!

By right away, I mean — once you have read the following list — you should decide which next step in self-education you will take, pick up the phone, order whatever next round of educational resources you can afford, set up three-ring binders with tabs to collect the trees and backup materials, and draft the initial and really rough handwritten versions of your TOC logic trees — all within the next 48 hours. If you have read this far, anything else is just procrastination.

1.   *Read or re-read* The Goal *and* It's Not Luck. If you have not read the 1992 version of *The Goal,* have not read it lately, or have not read it at least twice, I strongly suggest you do these things. Read this time with a pen in hand to make the notes that give rise to a thorough reading. Be sure to devote adequate time to both of *The Goal's* introductions (the 1992 version contains one introduction each from the original and the revised editions). Ditto for reading or re-reading *It's Not Luck.*

2.   *Decide which next step in self-education you should take and arrange to acquire the additional TOC educational resources.* If budget will allow, every company location should have a complete TOC "ready reference" resource area — books, audiotapes, videotapes, and educational programs from APICS or other authoritative sources. This is an area for which the principle of "penny wise and pound foolish" applies. If you possibly can, make the investment in self-education resources for your people. It will pay off in terms of greater support of the program, greater individual confidence and competence, shorter time required to benefit from your company's work with TOC, and lower amounts of external resources needed to succeed.

3.   *Set aside a three-ring binder with a set of tabs to support the project called "Introducing the TOC Management System".* This first binder will contain the evolving TOC logic trees and supporting materials used to plan and accomplish the successful introduction of the TOC management system within your organization.

4.   *Set aside another three-ring binder with tabs for the project called "Management of My Life".* Your efforts to introduce the TOC management system at work will take place in the context of everything else going on in your life. This second notebook will be used to focus your efforts to identify, balance, sequence, and accomplish the various objectives in your life. This second binder will hold your own private TOC logic trees and supporting information prepared from your individual point of view. It will support your own path of observation, planning, and action in your career and life.

5.  *Set aside another three-ring binder with tabs for that inevitable next project.* Keep this third binder handy because, inevitably, some other project will arise, either at work or at home, that will be of sufficient complexity and importance to warrant working it as a separate project. It *will* happen. You may as well have that third binder handy. For example, if you are a member of the board of directors of an APICS chapter, you can use it for general chapter management or for new initiatives to serve members, customers, or community. Maybe you are running the chapter, lodge, or specific service projects for the Rotary, Optimists, Moose lodge, Masonic lodge, Jaycees, Scouting, Little League baseball, youth soccer, drama workshop, Knights of Columbus, chamber of commerce, church, mosque, synagogue, or other organization of importance to you. Even closer to home, maybe something is prompting you to make changes in what you could call your family management system. I believe everybody has what you could call a "family management system" — a collection of views, roles, rules, and periodic procedures. Most of the time these systems are implicit. In other words, everybody knows what the rules are, but people rarely verbalize them. Still, the rules are there. Maybe you have arrived at the view that your family management system would likely benefit from a rethinking of habits, objectives, rules/policies, and priorities. This involves other people. You really care that everyone "wins" on a right and just set of fundamentals, You may think it's possible the next round of activity might get caught up in your own or someone else's human weaknesses or "blind spots". Therefore, you may decide to take a thoughtful and methodical approach to sorting things out and influencing things. In this case, you label the third notebook "The Family Management System Project".

6.  *Create "vision statement" future trees for each important project.* The first page of each binder should be a table of contents. The first tab should contain an executive summary and the latest version of the TOC future logic tree that summarizes the current "vision" or plan for the project. For examples of summary "vision" future trees, refer to the examples in Chapter 1 (Figure 1.1 and the *Mission Impossible* trees) and those in *It's Not Luck.* The notebooks can also contain plain text documents, presentation slides, spreadsheets, and other materials that support or explain the "vision statement" logic trees. Of course, if the "notebook" is really an individual, family, or corporate electronic intranet, the vision statement logic tree, as well as the

supporting information, can be stored in the form of various data objects connected via hypertext links. This type of electronic technology is not necessary to make effective use of TOC. Good old-fashioned notebook binders work just fine.

7.  *Continue to add to the TOC logic trees — your "learning organization knowledge base" — for each project and notebook.* For example, to support the project to introduce the TOC management system to your company, the future tree should depict the benefits you expect from introducing the various elements of the TOC management system. You can begin by adopting the generic future tree shown in Figure 1.1 as your first draft and adjusting it to the facts of your particular situation. You can capture facts about your situation on evolving drafts of current-reality and evaporating-cloud logic trees. As the plans become more detailed, you can move from the single-page summary future tree of Figure 1.1 to the number of future tree pages required to describe the plan in sufficient detail. You should then create prerequisites and transition trees to accomplish each injection on Figure 1.1. From there, you can use transition trees to document both existing and new business processes. Continue to do this for each project and notebook.

## Use This Book To Get Started Right Away

These recommendations are intended to help you use information in this book to get started right away. Many readers will find enough information is this book alone to get started with the TOC thinking processes in support of a first few obvious important projects.

The idea is to protect you from thinking that you need to wait until you receive more materials before you begin your work with the logic trees and other TOC management tools. No matter how many things you wait for, or how long you wait for them, the first steps to building logic trees will be the same. You should take those steps now, and not wait.

As you get started using this book, make sure you are taking the step of re-reading, reading, or gaining access to *It's Not Luck* and are making plans to gain access to other logic-tree thinking process educational resources from APICS or other authoritative sources. The right approach is to dive right in to use logic trees as you begin to line up sources of additional information. But *do* dive right in.

### *Trees by Hand or Through the Use of Systems? Yes.*

A tactical note: In this age of graphics, scanners, and document management software, one might argue that notebooks are "old hat". I am a proficient computer user and still find that handwritten trees are often the best tools. If you have a scanner, you can scan your handwritten trees into graphics files and attach them to multi-media documents anyway. If you prefer to have your notebook in the form of an "electronic book" with hypertext links and such, suit yourself. However, while I have access to and use these tools, I often still find it useful to print the trees for handy storage in a three-ring notebook. Suit yourself. The issue, after all, is the quality of the thinking and the effectiveness of the action, *not* the format of the presentation. Use the format that best supports your effectiveness. But, in any case, don't delay to … just do it!

## No Time Like the Present!

The concepts and processes of TOC enable people to understand their world better, and to know better how to take effective action. If we do not like the way things are turning out — or even if we do — we can get into the game, make a contribution, and make a difference. We already do make a difference, whether we know it or not. We may as well know it and act on it. Good luck!

Don't wait! Just do it! And have fun!

## What They're Saying Is True Enough To Matter

"In the quickness of our haste
It seems we forget how to live ...
Our individual roles we think
Not so important to the plot
The big picture unseen ...
It's the dawning of a new era.
People consciously don't care."

            ~ Gwen Stefani and Tony Kanal (of No Doubt),
               "World Go Round", on *Tragic Kingdom*


## What They're Saying Is Doable Enough To Do

"We've got to find another way
To make the world go 'round."

            ~ Gwen Stefani and Tony Kanal (of No Doubt),
               "World Go Round", on *Tragic Kingdom*


## So Let's Get to It!

"Just do it!"

        ~ Nike


## Thought at the Meridian

"There does exist for man, therefore, a way of acting and of thinking which is possible on the level of moderation to which he belongs. Every undertaking that is more ambitious than this proves to be contradictory. ... It is those who know how to rebel, at the appropriate moment, against history, who really advance its interests.

"No possible form of wisdom today can claim to do more. Rebellion indefatigably confronts evil, from which it can only derive a new impetus. ... Even by his greatest effort man can only propose to diminish arithmetically the sufferings of the world. But the injustice and the suffering of the world will remain and, no matter how limited they are, they will not cease to be an outrage. Dimitri Karamazov's cry of 'Why?' will continue to resound, art and rebellion will die only with the last man. ... Then we understand that rebellion cannot exist without a strange form of love. ... Real generosity toward the future lies in giving all to the present. ... All of us, among the ruins, are preparing a renaissance beyond the limits of nihilism. But few of us know it.

"Already, in fact, rebellion, without claiming to solve everything, can at least confront the problems."

        ~ Albert Camus, "Beyond Nihilism", in *The Rebel*

# Epilogue

## A Riddle, But Follow the Rules

Here is a little riddle for long-term thinking. Don't think about this one until a year has passed since your first reading of this book or since you last thought about the riddle, and only if you have been actively working with TOC principles and thinking processes (logic trees) in the year's interval. Here we go. Let's say certain particulars of a person's background, in the eyes of some observers, were insurmountable obstacles to the acceptance of that person's ideas, no matter how valid or useful the ideas. Let's also say that a change of perspective happens in the observer, a change that delivers an even deeper and more serious commitment to the observer's original basis for the objections. Finally, let's say the new perspective also converts those *same* troublesome aspects of background into the basis for viewing that same person as one of the perfect and necessary champions for the observer's original and cherished cause. *The challenge:* Under what circumstances could such things be true? Remember the rules — don't think about this for a year.

## Thank You for Taking the Time

This seems the time to stop and say,
Many thanks for stopping by today.
I hope you've found things useful here.
At a price in life that's not too dear.

I've listened well to write this book.
You've listened well to read it.
Who listens well is a very fine friend,
What's next, my friends? Let's do it.

*Good luck and God speed!*

# Notes

## Notes for the Introduction

1   If you don't like to read footnotes, this would be a very good time to muster up some really fierce determination *not* to read any of the many notes in this book. On the other hand, if you're that dashing and dynamic sort who shares my enthusiasm for reading dictionaries, telephone directories, and all the available footnotes, you're in for some real fun here. Enjoy!

2   I'm whitewashing a fence, and it's really fun![3]

3   Yes, you're reading this right. This is a footnote applied to a footnote, another in a really remarkable series of innovations from McMullen Associates. This particular note acknowledges the possibility that note #2 didn't work very well with international readers, which regrettably may be somewhat of a problem throughout this volume. I would ask my international readers to please consider this book to be, in part, an opportunity to add richness and color to their already very advanced skills with the English language. To explain, note #2 is drawn from Mark Twain's classic story of a boy and his friends, *Tom Sawyer.*

4   TOC is an improvement to the process of improvement. Therefore, it makes sense to apply it to our world outside of work as well.

5   There's something in this book for everyone to like. There is also something in this book for everyone to *not* like. And I'm not just splitting infinitives here. Having everybody both like it and not like it means the book is right on target. The world has always required views and tools for dealing with differences among people. Since this book is about tools that people of all styles, of all experience, can use to go beyond differences to a shared purpose, it (the book) had better have a little bit for everyone. It does.

6   This note concerns the middle quotation on the Introduction's facing page. I like Emerson a lot. His essays offer a great deal of wise insight, direction, and motivation. I find this particular passage to be very wise and heartwarming. (Thank you, Margaret, for bringing it to my attention.) I still have to chide the great man from Concord for not, well, *challenging* us a little more in this passage. For instance, suppose you leave only a small

garden patch to posterity? Should that be to have succeeded in an entire life? On the other hand, maybe that canny Transcendentalist is more clever than he seems. Ever notice how, when you think and feel that you are (a) a "success", and/or (b) a child of God, and/or (c) something along the lines of presumptively "okay" that you perform better at everything you do? You become more, well, *well?* Maybe *that's* it. Maybe Emerson foresaw that this quote should show up on the first page of this very book. Maybe he knew that a few people would consider starting slow with TOC, especially with the logic-tree thinking processes, for fear of struggling a little bit. So he established a measure of "success" that is doable and attractive and can serve as sufficient rationale — if we cannot come up with some other three thoughts — to think of and feel about ourselves as, well, a success. Then onward and upward from there! Gotta hand it to that Emerson. He's an incredibly smart guy. (Incidentally, Ayn Rand disdains Emerson in her technical/polemical works as a "little mind", an out-of-context, twisting, false-strawman slam on his clever "foolish consistency as hobgoblin of little minds and minor statesmen" quote from *Self Reliance*. She included this in her 1974 speech to the U.S. Military Academy graduating cadets at West Point in 1974. What is also funny is that she could only wish to have had such a useful impact on the world. Even if you accept her reasons for the harsh appraisal, it only demonstrates that doing something less perfect but in the right way (Emerson) is often better than doing something more perfect in a wrong way (Rand). Next problem, though, for Rand is that, while her system offers some good ideas, the flaws are also huge. This means, on average, it is not clear that her work is any more profound or perfect than Emerson's. It gets worse yet. Over time, good intentions and thoughtful process are more important, more helpful and more appreciated than the vicious sharp edge of isolated, over-stimulated, and over-reaching intellectuality. Sorry, Ayn. Better luck next lifetime. Or maybe you don't do those either?) One last late-arriving vote on the issue of success and small gardens: Voltaire's *Candide* sends us a "yes" vote for equating "success" with garden patches, even and especially in the case of small gardens.

7   I gave you fair warning there would be a lot of notes.

8   I should acknowledge the heavenly voice that came in with the breeze one Monday or Sunday morning. There is, well, no doubt but that it gave rise to much of the energy and inspiration needed to drive this little book home at the very end of its long journey. One never knows, does one, what future differences one might make when publishing something original and true into the public domain? One could say that, to the extent this little book helps find, well, another way to make the world go round, that that voice and its friends helped, too. While we are on the subject of key, late-in-the-game influences, I hope another person — a message returned to view at a time and in time to help me scare off some bogeyfolk and get some important things said my way and true — will think this little book is worth one of his "*wows*". If not, oh well; better anyway to get it right. Whoa! Thought this note was done days ago. Then, there I was, minding my own business, wondering about whether and how to adjust certain passages that might be too far out for some or too far in for others, and then — look out! —incoming! It's Cruise again, this time in *Jerry Maguire*, the movie about the sports agent and his unfathomably sweet and lovely assistant. The film suddenly — and, I guess, coincidentally — is hurtling in the direction of my VCR, via the good offices of my best friend. Scary. Gotta tell ya that writing this book has sometimes been as tough (but also sometimes as easy) as reading the scoreboards in a Kevin Costner movie. (That Costner puts those kinds of scoreboards into his movies and the movies have been popular tells us something interesting, doesn't it? I mean, there is some sort of "secret" in a lot of people's lives that's not really much of a

secret. If I'm not mistaken, that has ramifications of importance to managers in the area of perception and reality of possibility. Cutting that essay to the out-takes file for another book for another day. That is if this one ever gets done.) For those of you who haven't given up on this note, get this: Having written this note this far, on the day I am putting the final touches on the manuscript — trying to balance getting it perfect with getting the thing shipped to the publisher, still wondering and fussing about whether to include several passages sprinkled around the book — and having already shipped Chapter 6 with the little piece on "meaningful coincidence", don't I get Redfield '93 and '96 dropped back into my lap. Read both cover to cover this time; I had only skimmed the first one before, with the primary take-away the match of the nice "meaningful coincidence" idea with what life looks like a lot of the time. In Chapter 6, I acknowledged him but still poked a little gentle fun, but I can see now that he has anticipated and solved a lot of dangerous problems. Are his concepts "true" or "the truth"? As developed elsewhere, those are questions I cannot answer very readily. Now ask me if they are *valid*, in the sense of whether some of the concepts order a lot of data of everyday experience. The answer is, "Oh yeah, and how!" Thought you'd be interested to know.

9    Wanted to let you know that I did, in fact, finally succeed in working the subtitle of this introduction into the body of the text. Just tying up a few loose ends here in the final rounds of revisions to the book.

10   Goldratt, E.M., What is the Theory of Constraints?, in *1995 APICS Constraints Management Symposium Proceedings*, April 26–28, Phoenix, AZ, APICS, Falls Church, VA, 1995. Dr. Goldratt's presentation is recorded on a set of three audiotapes (APICS item #01503).

11   If Dr. Goldratt, in fact, has done this, he will surely someday win some sort of impressive prize. For what? For something like ... I don't know exactly ... but something like being the first or the best to articulate scientific thinking methods, specifically in a manner that allowed such a rapid and substantial expansion of their use that the increased use came to alter the weighted-average style ("weighted average" because, after all, almost nothing is accepted and adopted by everybody), of individual thinking and cooperative action, in all walks of life, all over the world. Or something like that. Again, I am not qualified to rule on how a peer review process would or should cause "real scientists" to weigh in on that sort of thing.

12   You, dear reader, have a vote in the degree to which Dr. Goldratt's contribution becomes a global standard and, thus, at least to some degree, becomes historic. There has been substantial impact already, even without conditions of professional literature, professional consensus, and packaging concerning TOC being correct for high growth. That means TOC and Dr. Goldratt are already "a little bit standard" and "a little bit historic", which cynics might say is like being "a little bit pregnant". This book is designed to break TOC's market constraint, which means, if it is going to work as intended, lots of readers like you will need to use TOC within your organizations and in various projects in your lives. The more organizations and individuals adding the TOC management system to their tool kits, the larger will be the degree to which TOC and Dr. Goldratt's contribution have become standard and one day "historic". As those great philosophers Bartles and Jaymes once said, dear reader, "Thank you for your support."

13   There are three additional reasons I am qualified to rule on the matter of which concepts and theories are relevant for making progress in the field.

   *One:* My professional background has been practical enough to include being a naval officer, shipboard missile guidance systems officer, officer in charge of nuclear reactor instrumentation and control systems, officer in charge of steam turbine generator electrical power generation and distribution systems, the "right full rudder" and "all ahead 1/3" type of shipdriver during restricted operations alongside aircraft carriers and while within narrow-channel navigation voyages, policymaker and on-site inspector for a major international oil company's supertanker operations (tugboated to 1/4-mile-long ships off-shore Texas, helicoptered into Persian Gulf, and iced over in the North Sea), computer systems reengineer, high-speed digital and voice information systems designer for large global corporations, field operations manager behind barbed wire fences and both outside and inside picket lines in beautiful downtown Newark, corporate labor relations advisor dealing with national-level grievances and arbitrations and federal court case preparations and contract negotiations and joint union/management programs to predict and mitigate future effects of technological change on the number of jobs and nature of work, reengineering project manager, initial failure as salesman, success as marketer, huge success as district sales manager, corporate financial analyst, corporate internal consultant, software company vice president, ex-software company vice president, board member for an urban community mental health center, and a lot more. It's not that I am saying I have done everything, or even that I know everything. What I am saying is that — while I may not be able to prevail in complicated and high-powered theoretical debates in the ivory towers and rubber rooms of the world's intelligentsia, over what's *really* a science, and what's *really* a theory, and what's *really* new and what's *really* not — I am here to tell you that I am qualified to comment on what sorts of ideas and programs make a difference in the field, in day-to-day operations.

*Two:* My educational background has been sufficiently both practical and theoretical to include electrical engineering and professional studies at a well-known military college (exactly which one is a secret for the moment — if I tell you which one right now, I have to, well, okay, there is an author's bio at the beginning of the book), an advanced degree program in business administration from a well-known university (another secret), and a program in nuclear management from a well-known and iconoclastic U.S. Navy figure who built this country's nuclear submarine force and influenced its commercial power generation systems (and whose initials are Admiral H. G. Rickover). In other words, I respect the fact that theory is not "just theoretical" when it serves practical purposes.

*Three:* The self-directed part of my experience and education over the years has been optimistic enough — and even idealistic enough — to include careful readings, consideration, and synthesis of works from pretty much the full range of competing world views, both mainstream and not. If you are likely to understand what I just said, you already do, so I won't elaborate. Except here's a pop quiz: What happens when you combine Bach's *Seagull, Illusions,* and *Running from Safety* with Pirsig's *Motorcycle,* Camus' *Rebel,* and TOC's structured thinking processes? Answer: Progress, but with far fewer side effects, despite historically high degrees of complexity and pressures, and with more people — from more and more different walks of life — intelligently understanding and helping to get things done.

So what's the point of these three additional reasons, other than to blow my own horn, sell my own experience, and wax pop literate in order to appear well-read and appeal to the widest possible consensus for cooperation? Is it to say my opinion is better than that of anyone else? No, it's just that, for some reason — and I prefer to believe there is some sort of reason — I have been led through a pretty unusual range of fairly extreme opposites of experience during my 44 years of life. Which means, dear reader, no matter where you are, or where you have been, or what you think, or why you think it, you and I most likely have things in common that motivate us to want to see this increasingly small world move in a constructive direction for our children and grandchildren (Oh, come on now. If you don't have children, then for your nieces, nephews, neighbors, and all the kids you see on TV. Let up on me already with this being literal. It's time to bring making the world a better place back in vogue!), family, friends, lovers, and, oh, by the way, ourselves. "Win-win", by definition, means you and I win, too.

14 Larry Gadd, the publisher of all three books, would be the one to know the true numbers, but it seems clear that, taken together, the three books have sold in the millions in over a dozen languages. To order any or all of them, contact the APICS Bookstore at (800) 444-2742, ext. 2350. Speaking of TOC reference materials published in languages other than English, even Mr. Gadd might not know of all the translations that have been done. Lisa Scheinkopf once said she had seen an unauthorized Malaysian version of *The Goal*. I recently had lunch with a manager in a large, well-known, brand-name Japanese company, a company recognizable both in consumer and industrial markets as a huge, powerful, and successful outfit, who has been translating TOC concepts into Japanese lately for use by his colleagues around the world. Dr. Goldratt, until recently, has said he was unwilling to assist in making translations of his TOC materials into Japanese due to a concern about balance of trade and power. This Japanese manager has been taking matters into his own hands. Stay tuned, folks. We keep it light-hearted here in order to have physics be fun and to keep us all awake, but there is a lot going here on that really matters.

15 One of the finest, most concise, and most effective self-motivation and self-development works ever written is a little book called *The Greatest Salesman in the World*, by Og Mandino. It is only a little bit about sales, by the way. It has a chapter devoted to the "do it now" theme. It has very clear cause-and-effect logic, sort of a family of text-based TOC future trees, on fundamental matters such as how bad habits are to be replaced with good habits, how to manifest practical and encompassing interpretations of the concepts of brotherly and neighborly love, how to inculcate a natural self-esteem, how to develop a unique and effective personal style, how to maintain a sense of humor, and more. This is good. However, other chapters — such as the ones on persistence, setting big objectives, and "doing it now" — can have the effect of sending oneself or other people unnecessarily forward into a large number of brick walls at very high velocities. In the absence of a more focused alternative, that is good advice and practice, because better to hit walls, learn, pick oneself back up, take on the next wall, and continue to accomplish and learn at least something along the way than to sit and rot. Like the rest of the book, these chapters are excellent and inspired stuff. The problem, the opportunity, is that they leave open the *huge* question of how to approach the three big questions — what to change, what to change to, and how to cause the change. Combine Og Mandino's *The Greatest Salesman in the World* with the TOC structured thinking processes and look out, world! There are many other combinations of TOC with traditional western world self-help technologies and literature that offer tremendous potential, but this one is a real gold mine. Note for Mr. Mandino: Here's another heart-opener/heartwarmer: "Whenever you see any adult, see him or her as the beautiful little girl or beautiful little boy he once was to someone and to himself." [English really *does* need some gender-neutral pronouns.] Anyway, try it. It's magic. Well, not *magic*, exactly. At least not the David Copperfield or Bill Yeats kind of magic. I just mean, it works. Here is how well it works. There are a few people who have taken advantage of my good nature in contract situations. This technique works so well it creates the illusion that there may even be some good in them! That means, for garden-variety irritations, this will be an extraordinarily effective technique.

[16] Hats off here, of course, to John Pirsig, author of that oft-quoted and rarely understood book, *Zen and the Art of Motorcycle Maintenance.* Mr. Pirsig was able to reach a wide audience with his demonstration that, if we understand the experience of "quality" a little differently, all of us very different attitudinal and experiential types (William James, Jung, and other prominent typologists are right this minute turning over in their graves) can stick to our stories about, on the one hand, spiritual/philosophic/aesthetic purity and, on the other hand, satisfying/moral/legitimate/practical work in the world (including the industrial world, by the way). We have not made best advantage yet of Pirsig's vision in the western world, as we are still fighting each other over the "either/or" of "have industrial organizations and technologies and property" or "have cultural conditions which nurture the spirit and better aspects of the character of woman and man", instead of making the "both" more manifest in life. (In other words, are the petroleum refineries and landfills and machine tool solvents not also, ultimately, the creations of Mother Nature, in whatever terms you prefer to think and speak about Her, Him, or Is?) TOC is the kind of tool that can dramatically accelerate the process of getting the benefits of the "both" without getting tangled up in historical entanglements, by digging deeper for shared purpose and shared intermediate objectives.

[17] I forgive you, Mr. Leadership Institute. The question is whether *you* can forgive you for failing to institute leadership, instead settling for stoking your own ego and brandishing your own superficial conference room stand-up comedy, by bashing the experienced senior managers of the firm, whose complex responsibilities precluded taking your overly simplistic path. (Note for all but one reader: There is an inside joke here, and I *do* mean inside joke.)

[18] We should think Joseph Schumpeter here, with his notion of the creative destruction of capital. What is true for our own and the competitor's capital is not necessarily true for our senior and experienced colleagues, at all levels of our organizations and everywhere else around us.

[19] Noreen, E., Mackey, J. T., and Smith, D., *The Theory of Constraints and Its Implications for Management Accounting,* Institute of Management Accountants (IMA) Foundation for Applied Research (FAR), Montvale, NJ, 1995. (APICS item #03356). Dr. Eric Noreen is nobody's fool himself, by the way. Among other qualifications, he is a member of the University of Washington business school faculty, a Price Waterhouse Professor of Management Information and Control at INSEAD, the European Institute of Business Administration at Fontainebleau, France, and the author of one of the leading university textbooks on management accounting. "But, then," say the die-hard, allocation-based, product costers, "what does *he* know?"

[20] Yes, I'm aware of the subjunctive thing here. Problem is I tried to use "... would they be building?" and it didn't really read right. "Are they building" adds a little immediacy to the tone, anyway; "... would they have already been building" is also a possibility, but that's getting a little ponderous. God forbid we would ever get ponderous! So, sorry about that. Couldn't be helped.

[21] "Good controversy" fuels constructive processes of institutional renewal. "Bad controversy" wastes time.

[22] I know, I know, I know already! These days, nobody has time to read in hotel rooms on typical business trips. My suggestion: Read faster on the plane.

[23] What's this doing here? Answer: The TOC thinking processes and logic trees help individuals and companies sort out what objectives to work on when situations are complicated. It is my own observation (don't blame it on TOC) that individuals and groups who

think and live in the context of an experience of "prayer" sometimes face the same challenge in knowing what to pray for. (Ever wonder, by the way, about exactly when, why, and how the everyday activities of thinking and planning cross some sort of a line and become a different thing called "prayer"? Interesting.)

24 The "we all" and "they all" derive from the fact that I am originally from the South. That's South Carney, near Parkville, in Maryland, which is near Towson, but a world apart. All of which is below Mason and Dixon's line, y'all.

25 Ever think about how the so-called "simple virtues" play out into large issues and events? A hostage situation developed in which officials pretty much had to do what they had to do once confronted with the problem. But, deep in the details of the media accounts was the story of how one of the terrorists involved got started in such a career. In essence, there was an instance of a child of a politically well-connected family in that country, who committed an unprovoked and horrible wrong to an adult in a less powerful family. In an instance of dirty justice, the privileged child got away with it. The children of the wronged adult were transformed and radicalized from living simple and quiet lives to pursuing lives that gave rise to them showing up as "terrorists" on television in the hostage situation. How much of the complex churn of politics and war is a result of a failure to take the "simple virtues" seriously? Talk about minimum number of simplest concepts with maximum utility. What if people in all walks of life got really flipping serious about them? (Again?) Don't get me wrong. I have no illusions that wrong-doing, radical politics, and war will go away completely in my lifetime. I also have no doubt that, within the extremes of no such problems and much such trouble, there is a very wide range that is, in large measure, influenced by many individual choices made by individuals. Your and my behavior toward some one other person may make the difference, today. (As one of the millions who were swept away during college days by the energy of *Atlas Shrugged*, I'm qualified to offer that Ayn Rand has not helped with this "simple virtues" matter, what with her incessant attacks on altruism. There are some legitimate points in her attacks, but they are too subtle to overcome the overpowering first impression she creates with this unfortunate choice of terms in her thinking system. It is also not helpful that her novels — especially her most popular, *Atlas Shrugged* — deliver characters designed to be attractive role models, but whose choices in life are always to run away from human nature, politics, and life. It is not enough that a thinking system be correct on some matters profound. It must also produce constructive practical effects. Same goes for Nietzsche.)

26 Common sense, of course, is not so simple and is not something to be taken for granted. In the late 1700s and early 1800s, a group of Scotsmen were stunned at what they considered to be the absurd outcomes reached by George Berkeley, John Locke, and other inventors of idea systems. I do not want to get into whether Berkeley's outcomes were actually absurd or whether they simply represented another view of life that also can be shown to explain and predict a lot of the data of experience. I also do not think it is useful here to get too far into the issues of how to break ties when choosing among concept systems that have one or another degree or scope of validity. The point here is that the Scotsmen (and probably Scotswomen) who were given voice by, among others, Thomas Reid, espoused a "doctrine of common sense". The doctrine asserted, in part, that the common error committed by the major idea systems of their day was that not enough emphasis had been placed on confirming the relevance of the ideas against the common sense that the average and unsophisticated person shares with the sophisticate. They felt that the task of the thinker was not to question and evolve the certainties within the

experience of large numbers of people, but rather to work at understanding the nature and origins of those certainties. Interesting criterion, but, in my estimation, a false choice. My guess is that the right answer is a bit of both. I'll bet a buck that either one without the other is doomed from the start to miss the proverbial boat.

# Notes for Chapter 1

1   Pop quiz #1. Food for thought: If an effective pilot's art for that stormy sea of meaningful coincidence called "life" were to appear in our lifetime, what would it look like, and why?

2   If you like this little dialogue and bookstore tour, you can give the credit to Drew Gierman. He told me the original version read like I was from Harvard or something. I did not ask him whether that was good or bad, but assumed it meant some of the tight, hard-hitting analytical prose that contained several points per line might want to be lightened up a little with some more lively prose that maybe took several lines to make a point. If you do not like it, blame it on me. To see how it read before, just ignore the little dialogue thing, the bookstore tour, and the "what's wrong with this picture", and read the other paragraphs. The difference in styles will be obvious. So, obviously, this is just a sneaky trick to get you to read the "Management Science" positioning statement again. Wouldn't hurt. There's a lot going on there.

3   Actually, the *Two-Minute Manager* really *is* a joke.

4   A few readers may be able to detect a trend here. Hey, I can be allowed a little mentor and institution admiration.

5   His name finally came back to me. It has been awhile since I read his stuff, and my library is a shambles from sifting back through everything to write this book. He is Regis McKenna, another very smart guy and a worthwhile read. Excellent exposition of cause-and-effect linkages in the area of marketing complicated product concepts. No TOC logic trees in his books, though. Maybe we'll see some in new revised editions ... Regis?

6   Okay, I confess I really like Guy Kawasaki's stuff. It's just that I've heard so many of my friends tease him for his enthusiasm that I couldn't pass on that line. Go get 'em, Guy! (You, too, Guy. You did what you could ...)

7   In fact, I know it costs a lot of money to buy all those books.

8   This is another shameless plug for one of my *alma maters*. Think the school will appreciate it? (Ever think about how to form the plural of "*alma mater*"?)

9   Him or her, his or her, himself or herself, and on and on. Now that we seem to have moved out of the old comfortable (to some) him/his-centric modes of expression, we are missing a few useful words in the English language. How did we get by for so long without words to express the gender-neutral thoughts, "him or her", "hers or his", "he or she", and "herself or himself"? It's awkward.

10  This, of course, is a huge assumption. As we try to make some really sound, fundamental, and lasting progress in this area of having organizations deal thoughtfully and effectively with individuals — and therefore have important indirect effects with their families, towns, counties, states, nations, and cultures (effects, by the way, that justify their positions in the various laws by which all of us just plain peoplefolk allow them to exist at all) — we are not burdened by a great deal of precedent. Do *not* go away, half of you. I saw you shake your head on that one. Stay with me. Continuing ... So that was the wise-mouth, flip, cynical, and snide view. The other view that is also true is the respectful view that goes something like this: The world keeps changing. Organizations,

such as governments, are only capable of operating within the constraints (!) that our laws and behavior place on them. Organizations today are able to do more and *do* do more (leave that one alone, please) that makes sense for human well-being than they were able to do (and did) generations ago. New possiblities are always at hand, in every age, in every generation. They can always do more. But, above all, organizations are populated with (!) *people*, for Pete's sake! (Anybody know where that "Pete's sake" phrase came from?) It is *people* (humans, all too human) running those organizations. Most of them are already doing the best they can, within their social and organizational constraints (yes, that was "constraints" — which we'll get to soon) without shooting themselves needlessly in the foot with useless gestures. More will be motivated to do better. So let's not argue over whether the cynical or respectful view is the "true" view. Both are true. The question is, "What are the next high-leverage increments of possible progress, and how do we get them done?" Hint: One of the biggest ones, one that's within reach right this minute, has the initials TVA. Stay tuned.

11  The "Bookstore Effect". Yet another new term coined! I love it. Who said physics can't be fun?

12  Miles, R. H., *Corporate Comeback: The Story of Renewal and Transformation at National Semiconductor*, Jossey-Bass, San Francisco, CA, 1997. It's not a huge mention, but credit to Mr. Miles for mentioning TOC at all. TOC has been just a little too hard to "get arms around", so most people just haven't mentioned it. Also, Miles was obviously comfortable enough with the contribution of his own methods to the success that he did not feel it necessary to bury the TOC contribution so deep that it wasn't visible. That is highly unusual, especially for a consultant. Still, the purpose of the book is to tell the story of the success of Miles' and Amelio's methods and not to relate how TOC was used to save capital expenditures amounting to nine digits (that's dollars) for an entire semiconductor so-called "front end" and to accelerate the availability of capacity for new product introduction, giving time-to-market benefits. For that part of the story, you will have to refer to Arjun Israni's presentation, in the *1995 APICS Constraints Management Symposium Proceedings*, April 26–28, Phoenix, AZ, APICS, Falls Church, VA, 1995. The proceedings and the National Semiconductor audiotape are APICS items #04216 and #01509, respectively.

13  I can't be sure whether it is really "none" or "few" because every time I start looking through the indices of books I end up being interested in things in them and end up buying more of the blasted things and — what's worse — taking the time to read them. That's a good thing, of course, but this book has got to get to the publisher. Right, Drew?

14  Same could be said for that other important guide through the minefield of modern improvement methods, *The Dilbert Principle*. C'mon, Scott, there is a *lot* to work with here. There is saving the world, a physics(!), middle-school kids outshining adults with a procedure for making common sense a common practice, another — yet another! — set of root-cause analytical tools, and much more.

15  Used to call these things "sacred cows", but I get less and less comfortable with that phrase. Those cows are, well, *sacred* to a lot of people.

16  This one's for you, George. I didn't miss or forget your question. This is a pretty good one-sentence statement of the central theory. Experience says that this presumption has a very wide range of validity.

17  This is important. Many people argue that there is too much science already. I think not. The problem is too much science in selected narrow niches, with too little science in the upper level and more general areas. More science of management can and should be used to adjust both the activities of the traditional sciences and our ways of experiencing them.

[18]   For purposes of practical applications of TOC and the TOC management system, to delve into the interesting and important questions of the nature, origin, and significance of intuition is to move off the point, waste time, and maybe unnecessarily offend important stakeholders. In other words, TOC and the TOC management system work with intuition systematically and rigorously but take it at face value.

[19]   There is this issue of how to "stack" the concepts, as far as which should be "on the top" and which "on the bottom". Doesn't matter. Performance of the system is the point. So, some companies will use TOC as their primary and foundational management system. Others will have some other system and use TOC in support of it. Business management is not the only arena in which it is sometimes tough to tell the players (and the concepts) without a program. Take something simple and well-known such as the Ten Commandments, for example. You can find them in the books of Exodus and Deuteronomy in the Old Testament of the Judeo-Christian Bible, as received. In the new Roman Catholic Catechism, they can be found organized under excerpts from the New Testament. They continue to provide pretty powerful insight and guidance, whether over or under other ways of looking at things.

[20]   If this "intrinsic order" thing sounds a little "flaky" to you, consider how non-flaky executives, non-flaky top athletes, and non-flaky wartime commanders know in advance that they will always find ways to prevail in new situations where most of what is going to happen is unknown. This confidence comes from knowing there are patterns beneath the appearance of many rapidly changing variables. These non-flaky folk intuitively know they will find these patterns. Some would quibble and say these achievers are creating and not finding, but we are not playing a word game here. We are pointing to a pattern in reality that is important to high levels of performance. So, in summary, all we are doing here with this "intrinsic order" thing is changing this everyday reality from an implicit knowing that is not often discussed to an explicit concept that is explicitly acknowledged and discussed. It just so happens that the name in TOC is the same name used for the same reality explicitly used in the non-flaky hard sciences. What is *really* flaky is for Mother Nature to have one view of what works while you are certain that your view of what is flaky works better. No, I've never been called flaky. That's not it. But I *have* seen some very promising concepts tossed into the cultural wastebasket because the consensus views at the time lined up on the wrong sides of the issues. Hey, it happens. Life goes on.

[21]   I would offer as how these two concepts might be *a priori*, except I don't read or write in Latin. And I "kant" read or write in German either, a language where *a priori* things are, apparently, really quite popular subjects for exposition.

[22]   Goldratt, E. M. and Fox, R. E., *The Race*, North River Press, Croton-on-Hudson, NY, 1986, page 18 (APICS item #03202).

[23]   The reason these definitions of the goal and money are used is very simple: They work. It is not a value judgment, but rather a concept that has been tested for validity and effectiveness over a wide range of situations.

[24]   If you are likely to be inventing other and internally consistent sets of definitions of systems, goals, and measurements for situations outside the range of validity of the set being developed here, you will also be able to tell from the wording of these paragraphs where to make the necessary adjustments.

[25]   Goldratt, E. M. and Cox, J., *The Goal: A Process of Ongoing Improvement*, North River Press, Croton-on-Hudson, NY, 1992, Chapter 8 (APICS item #03201).

[26]   Goldratt, E. M., *The Haystack Syndrome: Sifting Information from the Data Ocean*, North River Press, Croton-on-Hudson, NY, 1990 (APICS item #03125).

27 The former is named for Vilfredo Pareto (1848–1923), the Italian economist and sociologist. The latter has been stated in many ways, but the basic message is that 80% of the effects are produced by 20% of the causes.

28 Here is an example of the latter. Companies often spend a lot of time and money and consume precious time that could be applied to other projects in calculating product cost numbers based on allocating "overhead" and other periodic operating expenses to specific products or services. Having gone to such great trouble and sacrifice, they then proceed to use them to make wrong decisions which harm the company. Don't laugh. It's not bad enough that jobs are lost with this kind of thing. Think about what people in companies have to teach young people who ask about the world of work (e.g., "Dumb things get done, but you will need to play the game. They confuse themselves with the numbers and you had best not get in the way. They'll think it is *you* who don't understand."). What is the effect on respect for leadership and authority? It is not fashionable yet to talk about all these separate subjects at the same time, but that does not mean that they are not directly connected. It has not worked — and never *will* work! — to make believe these things are not mutually influencing and to try to deal with them separately. Which is all to say, management accounting and corporate finance are much too important to be left to the professional management accountants and to the experts in corporate finance. More of us amateurs, who aren't smart and sophisticated enough to know that these are separate matters, must begin to pay much more attention to this. By the way, when jobs are lost, so are still other things in the society. The jobs lost by this are a partial cause of the disillusionment that influences the world you and I live in. Here is how it works. We all know this. Let's say a hard-working and loyal employee, who also happens to be a father or mother, tells a son or daughter that working hard and aiming to satisfy — or even impress — some legitimate authority is a good strategy in life. Consider what happens when that child then watches the parent look like a fool for having believed in what turns out to be such silly notions. There has been some focus on the effects on the employees themselves, but now it is time to spend some time understanding the effects on the children, their thinking, and their attitudes toward society. For what purpose? Not for hand-wringing, but rather for developing a determination to have businesses do a better job at this, partly via strongly encouraging things such as TOC's growth-oriented strategy approaches, TOC's thinking tools for building more intrepid growth-accomplishing operations, and TOC's TVA management accounting to make better growth decisions easier and faster. Hmmm ... management accounting. That's how we got here. Back to the text.

29 This is not a classical TOC principle. It's a new one. I like it. It's pretty valid, doesn't your intuition agree?

30 Okay, I confess. The truth is I've really kind of fallen in love with this particular phrase and have, quite honestly, been trying to sort of jam it into the book somewhere (anywhere!) partly to anchor a footnote, but with no success so far. One approach I tried was, "Time is the prime constraint, especially for people who are very busy." It's true and everything, but it leans just a smidge in the direction of the tautology, if you know what I mean. If you can think of better ways to work this phrase into the TOC body of knowledge, I would really appreciate knowing of them.

31 I do not really have much more to say about this particular catchy subtitle, except to tell you that it really appeals to me. Over the years, I have discovered that it is best when read out loud using a deep voice like Bill Cosby[32] used when intoning the word "Noah". Try it for yourself. Really. "Time is the prime constraint." And now once again. "Time is the

prime constraint." Chances are this exercise — and you — will make a very strong impression on the people around you.

[32]    Bill Cosby. American hero. What a powerful and important contribution that man has made. Showing us all how. Not only his humor brings tears to my eyes.

[33]    The problem with this is that once we busy people have something that gives us more results for less time, we start using the new spare time to get even more results. This, of course, puts us right back to the situation where, well, *time is the prime constraint.* Still, there may be something there that is good. Meanwhile, since I was not able to get "time is the prime constraint" into this book anywhere, I suppose I should just move on to the next topic. Thanks for listening, though.

[34]    Oh, great. We're answering the question with another question. Classic TOC.

[35]    You say you don't remember establishing these things? Gotcha! You skipped this book's introduction! You really should go back and read it. After reading the fine print in these footnotes, you will find the larger type in the introduction to be easier on the eyes.

[36]    I am trying to persuade the publisher to offer this book with a coupon for a discounted price for the video for use in logic-tree case studies and exercises. Just think of the possibilities. For example, "Create the situation-analysis and scenario-testing logic trees from the point of view of both Max and Ethan during the scene where Ethan is in the chair blindfolded."

[37]    Strange, but true, is that many who have been considered, due to the needs and deeds of the past, "highly successful" people lack these basic skills of human being. That's strange, or at least unnecessary, since every human being comes with these powers pre-installed. As with all other standard issue powers, these muscles increase in strength with use, and vice versa. Most people can locate this particular capability just north of the solar plexus. Others should look just to the right of the Izod logo. No, not *that* heart. The *other* one. The one that just feels like it's there. That this listening capability comes standard was amply and powerfully demonstrated outside the mainstream in the 1970s by Perls, Johnson, Erhard, and others. In the 1980s, these modes of listening, communication, and relationship rapidly became the mainstream, not only in therapy-like environments, but also in industry. Many companies and some entire industries adopted these simple, powerful, and "for-free" skills as standard, using policy, process, and culture to communicate and encourage them. Companies did this due to the obvious and inevitable changes in the global workplace, competitive, and life circumstances and not *only* because of babyboomer demographics and "liberal parenting". That's important. These new listening, communication, and relationship skills were, of course, not so new.

Consider the type of listening that has gone on in the confessionals of several theological traditions all over the world for nearly two thousand years. The leaders of those traditions could have designed that activity to be *sans* another person, but they didn't. Why? For the same reason the Perls, Johnson, Erhard, and other similar versions weren't. Because it worked to have someone on the other side listening at a level that empowered. Yes, I am aware that there is more that is important and is to be respected going on in the actual sacrament. The issue here is that there is an uncommon level of listening skill that can, and should, become more common. (Like most other skills, the trick will be for someone to verbalize the means by which gifted possessors of the skills do it, discern the common themes, and share the method with very large numbers of others in packages that don't unnecessarily offend.) Speaking of packaging getting in the way of making useful insights and capabilities more common, if you recognized and associated some of the names I listed earlier in this endnote with ideas that were either "new" or "way out", there are two things you may be interested in doing. The first we have already just done: considered the nature and huge benefits to human well-being of the confession sacrament in certain popular spiritual traditions. The second is to read *Spiritual Exercises*, a little book written by a gentleman named Ignatius Loyola, who lived and worked in the 1500s and whose ideas and followers have enjoyed various locations within and outside the "mainstream" over the generations. Do these two things, compare them with the core themes of all that "new stuff" of the 1970s and 1980s, and you'll almost begin to wonder if anything is ever really new under this sun. What does it all mean? For one thing, some of the controversial names listed above have made substantial and courageous contributions to many people's understanding of the relationship between concept and experience.

For example, I'll never forget the experience, in April 1980, in my dorm room at Galatin Hall at Harvard Business School, when the combined insights from all those "way out" characters were still new. I picked up my copy of the previously nearly impenetrable Spinoza's *Ethics*, opened it to pages at random, and suddenly realized that I was now *understanding* what I was reading. I thought I had "understood" it before. But now I was even understanding the concept of understanding differently. For another thing, some of these people sometimes packaged their simple, universal, powerful, constructive, and important themes in other concepts and language and forms that were, in some respects — and, at least arguably — very poorly chosen. Fine. Granted. But, hey, think about it! The history of anything — and especially the history of ideas — is really a lot messier than we perceive it from the summaries we get of it in our primary schooling. One last time, what does it all mean? Answer: This is exactly the type of situation where the TOC negative branch process shines. Leave us not shoot ourselves in the foot by ignoring powerful and important old or new ideas simply because they showed up in a manner that produced, along with the desirable effects which constituted their promise (the trunk of the TOC future-reality logic tree), some undesirable effects of its own. To fix the problem, add some additional ideas, some new injections. Keep doing that to trim all the negative branches. In the end, we've secured, and not ignored, the benefits. How did we get here? Right, it's strange, but true, that many successful people aren't empowering listeners. But, from another view, it's not so strange, for two reasons.

Reason number one for "not strange": For one thing, not to listen to others at the level of empowerment can be a negotiating choice, and every executive knows it, at least intuitively. (For executives, it might even be a legitimate experiential and relationship choice, since the challenge of creating constructive change in complex institutions requires all the tools of influence one can muster. And, by the way, if you're not an executive, next time you think executives have an easy life, at least when it comes to selecting how much idealism they can build into their plans without stupidly shooting themselves in the foot, think again. Like you, the executive is a "people", too. Want better institutions? Then think often of these people as wanting to do it better and help to create circumstances in which they can.) Not to be effectively listened to disempowers. Empowerment and disempowerment, like hammers and saws, are tools, neither good nor bad. Use them for good (as best you can understand it) — or lose them. Reason number two for "not strange": Think of what the world was like before it became less small and less visible. Big strong huns created waves of power through rival kingdoms in one age, and rival industries in a later age. It was enough to brook no opposition, to have charisma and other Oberman powers melt opposition and create institution, and then have subsequent generations forget the ragged edges and convert a flawed human to spiritual icon. Even Nietzsche's ideas work in *that* environment, and without very aggressive trimming of negative branches. But not today, not in *this* environment. (Shame on Nietzsche for using his gifts and his ability to create phrases that can lift you off your chair, to predict and create a world that's *Beyond Good and Evil*, without providing some practical alternatives for dealing with a world beyond. Thus spake McMullen.)

38  It's the same kind of attention that the "time is the prime constraint" exercise, described above, can provide. Men and women in very nice white coats …

39  He has to be careful with this. Suppose he gets so good at it that he experiences himself as others experiencing themselves? Now *that* would be communication! (The phrases of the second sentence are *Knots* only a Laing could love?)

40  Which is like *Tool Time*, only different. Which is the only way I could think of working Al's partner, Tim Allen, into my book. Hey, I adore the show! Besides, any guy who can attract a wife like Al's partner has on *Home Improvement* deserves a very honorable mention. Besides, again imagine a book with endnotes that *all* read like Tim trying to recite his neighbor's words of wisdom. Hard to imagine such a book? No? Uh oh.

41  I would like all my readers to know something about the name of the process I sometimes call the evaporating-cloud logic tree. Dr. Goldratt once told me he wants the world to know that he named the evaporating cloud to honor author Richard Bach. Dr. Goldratt has described Bach as one of the three people who have had the biggest impact on his life. The term "evaporating cloud" refers to a scene in Bach's book, *Illusions*. The reference from the TOC evaporating cloud to *Illusions* is also documented in the footnotes of the *Theory of Constraints Journal* series. So all my readers know this now. I've done my duty. With the usual respect to Dr. Goldratt, I continue to use other names for the same process for two reasons. One, while I, too, have found Bach's works to be remarkable in their ability to trigger insights, which in turn have enabled me to see the world in a larger number of sometimes surprisingly valid ways, I don't think all of his ideas are always packaged and organized for practical use by large numbers of people, in all walks of life, all over the world. I've suggested as much, and respectfully, to Bach in some fan mail correspondence, which he has graciously acknowledged. The radio-show scene in *Illusions*, as well as telling slices of his other works, indicate he is aware of this packaging thing. Yet, for some reason, he seems to think it is constructive to use a style that some people find irreverent anyway. People may find it disrespectful, but Bach is not a irreverent guy who doesn't care. Quite the contrary. The signs are clear.

So what is going on? I have half the answer. To an extent, I can see what he is doing and why, because many people report becoming more deeply committed to the traditions in which they were raised once they have used Bach's unique and sometimes pretty risky out-of-context strategies for triggering important insights and experience. These new viewpoints, in turn, produce deeper understanding and usability of lessons that were presented originally in the context of the traditions of youth. (Yep, "usability" is a word. Just had my spellchecker check it. On the other hand, it did not seem to like "usability's". I guess it has never seen "usability" used as a noun in the possessive case. Just goes to show that even spellcheckers can only really understand words, concepts, and other symbols that they can attach to underlying prior experiences.) Also, credit where credit is due. As Bach's work triggers insights that end up causing very large numbers of very different people to appreciate, participate in, and support the traditions of the training of their youth, his work makes it very clear that he does not want anyone who gets all inspired and enthusiastic about his work to form some sort of new organized religion. Not a bad move for one of few guys on Earth with the juice to do exactly that, if he wanted to. There's more to Bach, but I don't understand it all yet. Though additional portions of his work become clear to me every month or so — sometimes in avalanches, sometimes in gentler insights, sometimes while running for coffee (vs. from safety), sometimes while cursing the latest leaky fitting in some blasted *new* garden hose — there are some of his passages that still lose me completely. Maybe I'll understand them later. Maybe they're wrong. Two, all these things considered, experience tells me that the same logic-tree process sometimes goes over better with a different name. Different strokes for different folks. As a result, I sometimes prefer to use the terms dilemma tree, dilemma logic tree, conflict-resolution diagram, conflict logic tree, and — very very recently — evaporating-cloud logic tree (ECT or ECLT, depending on whether one prefers three-digit or four-digit area codes). These terms were created by, as far as I know, Chesapeake Consulting, McMullen Associates, TOC Center and Goal Systems International (I think TOC Center did it first, and GSI wrote it down first in hardcover), McMullen Associates, and McMullen Associates, respectively. But, then, who's counting?

42 ☺

43 That, by the way, is what always happens when a concept closely approaches what is going on in reality.

44 That's as in *Tool Time*, only different. Again, ust wanted to get the big guy in here again. Who? The big guy. You know, Al's partner. (Sorry, Tim.) Just wish I could get *Funniest Home Videos* in here somewhere. That's going to be tough ...

45 Physically impossible. Remember this was 1980–1981, in the age of manual typewriters. I know, you are trying to forget. Me, too. Love these word processors!

46 That's a hint.

47 As discussed in Chapter 5, software systems will be developed to support large-scale and distributed-responsibility applications of TOC logic-tree diagrams. For those systems, naming conventions for logic-tree entities will need to be developed. For example, an entity name might have the format, *tbm970630-001aa*, to deal with the issues, respectively, of owner of the tree page, date that the tree page was created, entity number on that tree page, and version number for the entity. (That is if the best data architecture for TOC logic trees turns out to include the concept of a tree page.) But these things need not concern us as we get started with logic trees prepared by individuals or even by fairly large groups from within or among organizations. Lots of work has been done, and lots more will be done, using trees running from just a few to dozens to hundreds of pages, using hand-drawn trees and trees prepared with graphics programs of low to moderate power — in this era before the obvious families of comprehensive TOC logic-tree software support tools are created.

48 Nevermind.

49 The technical name in TOC for an idea produced by the evaporating-cloud process, for use immediately in life or in the analysis of a potential plan using a future tree process is "injection". But, the way things were developing in that example, I thought it might be a little inappropriate to be quite so precise. I wouldn't want anyone accidentally thinking about drugs and rock-and-roll, either.

50 Okay, okay, okay, already, Eli. I *said* "okay!" I am changing all the titles of the figures in all my chapters and papers and presentations back to "evaporating cloud". I can't decide which of these others is best either. The point is the process used to examine assumptions and generate ways to move forward. Right. I know. I got it. Yup. Sure. I *said* "okay!" What about "evaporating-cloud logic tree"? Hmmm ...

51   Don't you just love it when someone says something like "TOC's CLR"? Oooooh. Sends shivers ...

52   Though it may be hard to believe, this is *not* one of the formal elements of the TOC lexicon. It is just that the two cause-and-effect links in Ethan's tree were BS. If he weren't so effective, I'd be angry with him for not taking the procedural details of the logic-tree thinking processes more seriously sooner.

53   ☺

54   Some people such as Bobby Miller, one of my 47 roommates in those luxury quarters during the U.S. Naval Academy "youngster" (sophomore) training cruise in the summer of 1971. We played 500 Rummy all the way across the Atlantic Ocean, letting the score build to way out of sight. And didn't he keep ... making up ... new rules ... all the way ... across the water! The only thing worse about that cruise was when John "Mugsy" McGraw, already as a freshman (plebe) the brigade's (entire Academy's) boxing champion — and future Navy Seal — decides he wants to "spar" with Tommy "4.0" McMullen", the (at the time and for a while longer) collegiate lacrosse star (I didn't give me the name; the ubiquitous *they* gave me the name), down in whatever the damp lower deck was called that was used on this ship called an LPD-1 (Landing Port Dock, first one ever made), the U.S.S. Raleigh, to launch and recover amphibious assault (think D-Day, Iwo Jima, and the like) vehicles. It wasn't a very smart idea on his part, since I was no trained boxer. It was also really stupid on my part — again, since I was no trained boxer. John taught me to keep my left up by showing me exactly and repeatedly why that left is supposed to be up there. The nose is still more than a little crooked to this day, thank you very much, John, wherever you are. Next time, we make *you* a goalie and I give you a little taste of the smoke from my lefthand, underhand, ya-can't-stop-what-ya-can't-see rocket to selected locations on your now-we'll-see-how-hard-that-rock-hard-granite-body-really-is musculature. The cruise was not a complete loss. After all, there was Portugal. Was that beach, so lovely, called Estoril? Or was it something like Mah? Sintra? Yes, but no. It's been awhile, but delightful memories remain.

55   Other TOC experts use the terms "wet" and "dry" for what I call "loose" and "tight" trees. Each to his or her taste. I like tight logic tree, as in tight logic. That's partly because I don't even want to tell you where the original term "wet" came from. Sometimes you have to allow for the things that can happen when things are invented in one language and are translated into one or, in the case of TOC, dozens of other languages.

56   Oh dear. Fruitless future trees. I didn't do that one on purpose. It just flowed out of the keyboard just now. Sorry. Maybe this should go into the next edition of the APICS dictionary, as a standard TOC logic-tree thinking process portion, of the public domain TOC portion, of the APICS body of knowledge? Jim? Johnny? C'mon, guys ...

57   The intrepid, by reading the lines and reading between the lines, will use this book to get started right away with TOC logic-tree applications. If you're taking that no-frills, no-nonsense approach, get Goldratt's *It's Not Luck* (APICS item #03291) and Dettmer's logic-tree book (APICS item #03516) right away.

58   Technical term.

59   Another technical term.

60   He would be like Abe Lincoln, who reportedly told a colleague how he would proceed if given 30 minutes to chop down a tree. He'd use the first 25 minutes to sharpen the ax.

61   Alas, one of the all-time bad news/(good news) heart-throb stories in moviedom occurs when the one woman in the world who can get the pronunciation right, when saying

"Geev mih theh MAAAHN-ah, AAY-than" three times fast or even one time beautifully turns out, during that scene in the baggage car, on the London-to-Paris chunnel train, to be (and have been) a bad girl. Ah, well. I guess that's Hollywood. And life.

62 Noreen, E., Mackey, J. T., and Smith, D., *The Theory of Constraints and Its Implications for Management Accounting*, North River Press, Great Barrington, MA, 1995, page 91 (APICS item #03356).

63 Panel discussion involving representatives from Boeing Corporation, Printronics Corporation, Chesapeake Consulting, and TOC for Education, in *1997 APICS Constraints Management Symposium Proceedings*, April 17–19, Denver, CO, APICS, Falls Church, VA, 1997. The proceedings and session audiotape are carried as APICS items #04220 and #01576, respectively.

64 Altshuller, G. (H. Altov.), *And Suddenly the Inventor Appeared: TRIZ, the Theory of Inventive Problem Solving*, Technical Innovation Center, Worcester, MA, 1996 (Lev Schulyak, trans.).

65 I know, because I was reading it that night at dinner, D. Yes, I did see you, but I felt, given your dinner partners, that it was best not to pay you a visit right then.

66 Goldratt, E.M., Overview of TOC applications for industry, in *1996 APICS Constraints Management Symposium Proceedings*, April 17–19, Detroit, MI, APICS, Falls Church, VA, 1996 (proceedings, APICS item #04218; two audiotapes, APICS item #01535).

67 Goldratt, E.M., A TOC approach to organizational empowerment: correcting misalignments between responsibility and authority, *APICS, The Performance Advantage*, 7(4), 45–47, 1997. This edition of the APICS magazine also contains the article, "The TOC Management System for Manufacturers", by T. B. McMullen, Jr. (see pages 52 to 56).

68 You haven't really lived until you've run the gamut to qualify to supervise and/or operate one of U.S. Navy Admiral Hyman G. Rickover's naval nuclear reactor plants. Part of the process is reading, remembering, relating, and recalling the multiply-inter-referenced contents — reactor theory, engineering principles, and detailed operating processes — of your particular reactor plant's operations manual. There is intermingling of theory, science, procedure, rationale for procedure, detailed hands-on experience at your own and every other team member's reactor job stations, descriptions of plant sensors, the assumptions and theories underlying each sensor, and the range of plant operating temperature and pressure conditions over which each sensor is really saying what it appears to be saying. There is the constant admonition to know what assumptions you are making, to take nothing for granted, and, therefore, to be able, in an unusual situation, to construct a view of what's really going on, when the normal indications are steering you wrong. To think, think, and think some more as the action proceeds. Ask anyone who went through the "Rickover program" in the "Rickover era". They will tell you that it was *different*. On the other hand, if you have read Augustine and Aquinas and the systems of mutually referencing source materials surrounding the new *Catholic Catechism* ... or if you have explained to yourself how Torah gets you to Mishnah (with or without help from a Tanna or an Amora of one of the Gemara, and with or without the strategies of a Rashi, a Rashi relative, a Maimonides, or even, if you prefer altering the approach a little, a little help from helpers in Kaplan's or Scholem's area of expertise) ... or if you've taken Eli Goldratt at his word and rederived all of TOC for yourself, from first principles, as if Goldratt had never done it first (which is the right way to do it, by the way) — then you already have the flavor for the Rickover reactor program experience. Like I said, it's *different*. (I had wanted to express the *Qur'an*-based, *Mahabarata/Ramayana*-based, and

Shinto and Confucian and Buddhist and Taoist and Humanist and Animist bodies of literature and experience into equivalents of literature, study, experience, and law, but I wasn't as far along with them at publication time. There is no intent to exclude these and other old and new friends who may prefer to look at things a little differently but who still are interested in always taking some step together to make the world either a little bit, or even a lot, better. It may take me just a little bit more time to get my arms around some of these things, at least to the extent of being able to express them from the right perspectives. Gives me something to do in my spare time.)

[69]  A welded tubing company once walked away from several million dollars per year of additional profit because it violated the "good idea" half of this little rule from the Rickover Navy. This is also a beautiful case of managers confused by cost accounting.

[70]  Another technical term. No offense to Disney or anyone else with an interest in ducks. Certainly wouldn't want that irrepressible punster, Bill G., to suspect foul play.

[71]  Kingman, O., Delta Airlines, in *1996 APICS Constraints Management Symposium Proceedings*, April 17–19, Dearborn, MI, APICS, Falls Church, VA, 1996 (proceedings, APICS item #04218; audiotapes, APICS item #01548).

[72]  Noreen, E., Mackey, J.T., and Smith, D., *The Theory of Constraints and Its Implications for Management Accounting*, North River Press, Great Barrington, MA, 1995 (APICS item #03356).

[73]  "Throughput" as a management accounting term describing a money flow appears in the literature as early as 1983 in Dr. Goldratt's well-known paper, "Cost Accounting: Public Enemy Number One of Productivity" (*APICS Selected Readings in Constraints Management*, APICS item #05021). It appears again in the several editions of *The Goal* and *The Race* between 1984 and 1992. More detailed discussions of applying the definitions from *The Goal* were made in *The Theory of Constraints Journal (TOCJ)*, published by the Goldratt Institute in the late 1980s and 1990. The best source and one that is easiest to find is Goldratt's *The Haystack Syndrome: Sifting Information from the Data Ocean* (APICS item #03125). The discussions in Chapters 1 to 6 of Part I are easier to find than the *TOCJ*, are followed by additional very useful material, and are a little better than the *TOCJ* on this topic anyway.

[74]  I once heard an accountant (I think it was TOC accounting expert, Dr. John Caspari) joke that there are only three kinds of accountants: those who can count and those who can't. I think he might be right. He gives a very lively presentation, by the way. An audiotape of his work is available in the *1996 APICS Constraints Management Symposium Proceedings* audiotapes.

[75]  *The Goal*, Chapter 8.

[76]  Goldratt, E. M., *The Haystack Syndrome: Sifting Information from the Data Ocean*, North River Press, Great Barrington, MA, 1990 (APICS item #03125).

[77]  For example, the uses in manufacturing and remanufacturing operations in support of incremental scheduling and of decisions about whether to replace existing TOC perpetual schedules with new schedules. See McMullen, Jr., T.B., What a difference perpetual schedules, incremental scheduling, and intrinsic full pegging make to remanufacturers! (all manufacturers, really), in *1997 APICS Remanufacturing Symposium Proceedings*, May, Dearborn, MI, APICS, Falls Church, VA, 1997. There was a limited printing of the 1997 Reman proceedings, so the paper also appears in the *1997 APICS International Conference Proceeding* (APICS item # 04017). A full discussion will be provided also in the forthcoming McMullen, Jr., T.B., *TOC Manufacturing Systems*, a 1998 selection in the APICS/CRC-St. Lucie Press TOC series.

78 APICS, the Educational Society for Resource Management, Falls Church, VA. Information about APICS appears at various places in this book, including the section entitled "About APICS".

# Notes for Chapter 2

1  A book about physics, even one about physics for the rest of us, can be a little dull. So, I thought I would spice it up with one of those ethereal contradictions that lend an air of the mystical while introducing the element of suspense. How do you like it so far? I was thinking also of coining a Sixth Discipline, maybe even an Eighth Habit of Highly Successful People, a Science of the Long View, or even Thinking Processes of Attila the Hun.
2  Pronounced "Ellie", as in short for Eleanor.
3  Dinner conversation with Dr. Eli Goldratt in Denver, February 1996, during a stop on the APICS Constraints Management SIG/Goldratt Institute Senior Management Tour, "Overview of TOC for Industry".
4  To hear Eli's story of TOC's evolution in his own words and voice, refer to the audiotape from the *1995 APICS Constraints Management Symposium Proceedings*. The 4-hour, three-tape set is available from APICS as item #01503. As an extra added attraction, my own and Lisa Scheinkopf's dulcet tones are also to be heard on Tape 1. I open the symposium and introduce Lisa. Lisa introduces Eli. And that's still not all.
5  The present author. Did you guess it?
6  Noreen, E., Mackey, J.T., and Smith, D., *The Theory of Constraints and Its Implications for Management Accounting*, North River Press, Great Barrington, MA, 1995, page 149 (APICS item #03356).
7  I know what some of you are thinking, but I don't want to go down the road of the solution being created because someone set out to find it and not simply discovered it. It may matter in rarefied thought and conversation, but it just doesn't matter for practical purposes to guys and gals like me in the field. If we must deal with this at all, I'll grant that it is conceivable that there is even more potential utility in the notion of a created intrinsic order than exists even in the really very powerful notion of a discovered intrinsic order implied in this volume. But let's walk before we run, shall we? And let's not ask for trouble, shall we? Let's milk this discovered intrinsic order concept for awhile for its maximum constructive effect on well-being. Can't make things too perfect too quickly, can we? Got to leave some contribution left for our children and grandchildren to make. Besides, though I personally do not prefer or require to do it, you can certainly jump sides of this street easily if you really think you want to or must. "Created" and "discovered" are only concepts themselves, after all. When a new solution is in hand, it's in hand. Once it's there — even with very precise definitions of "creation" and "discovery" also in hand — still can't really tell enough from memory or other stories of origin to know for sure whether "creation" or "discovery" is more correct in any given instance. *What* did he just say?! Not sure, but I think it was "nevermind". Right, nevermind.
8  See, for example, Sheldrake, Rupert, *A New Science of Life: The Hypothesis for Formative Causation*, Blonde & Briggs, London, 1985.
9  Goldratt, E.M., *What Is This Thing Called the Theory of Constraints and How Should It Be Implemented?*, North River Press, Croton-on-Hudson, NY, 1990 (APICS item # 03341).
10  Dr. Goldratt's published works typically do not acknowledge the contributions of anyone else. Where they come close, they use words that could be interpreted either as faint praise

or as criticism. This, I think, is not unusual for someone in a professor/graduate student relationship with a great many graduate students. The originator of an idea wants all the credit; the graduate students want some credit, too — sometimes more than they deserve, since often they are re-inventing things the professor has already invented. These things are not unusual but are matters of degree and interpretation. I believe Dr. Goldratt takes this to an extreme, which is to say that, along with his many strengths which I admire, he also has a few faults. Why should he be different? On the other hand, this particular fault has slowed TOC's introduction and acceptance. On yet another hand, it has slowed things to a pace he could reasonably control to build the foundation of a legitimate management science. There are pros and cons here. In a situation like this, we have the luxury now of looking at things in retrospect. Where the success now seems it was always inevitable, it most likely didn't always look quite so inevitable to Eli. He had to call 'em like he saw 'em at the time.

[11]  A reference to *Grinding It Out*, a book by McDonald's Corporation founder Ray Kroc.

[12]  To be sure, that has happened during the early years of TOC, too. It will also happen sometimes in the future. The point is that the TOC system, with its logic trees, which include the "negative branch" processing procedure as part of the basic tool kit, has the intention — the determination — to avoid unnecessary calamities.

[13]  Speaking of ideas giving rise to things, here is one of my all-time favorite book titles: *The World as Will and Idea*, by Schopenhauer. Just think about what that title is saying. Is his view right? Right? Well, how about we shift the discussion from whether it's "right" to whether it's "valid"? I can deal with the question now. Yes, indeed, Schopenhauer's view certainly is valid in that it can be used as one way to order, explain, and predict quite a bit of what one experiences in day-to-day life.

[14]  Speaking of genius, here is another pop quiz, an exercise. The challenge is to build an analysis logic tree, a cause-and-effect diagram, that explains how the entity at the bottom of the new tree diagram — "Pat Boone has a new wardrobe" — leads inevitably to the entity at the top of the diagram — "Therefore, Pat Boone is a genius". I know this can come across as sort of funny at first. But, actually, I'm not kidding. Hint: Another entity is "Nice people come in packages that surprise some people, and that's dangerous to both groups". Mr. Boone is brilliant to see the problem. Brilliant to recognize that almost no one in the world could make the point as clearly as he. Brilliant to figure how to get the message visually through the pre-occupation of a busy society. And brilliant to be enough of a sport to just up and do it. Bravo, Mr. Boone, bravo! (Sure hope that's the reason.)

# Notes for Chapter 3

[1]  Don't get me wrong. By now, you know I like Emerson, but I have to quarrel with the great man on a few small points concerning this particular quotation. This sentence gives the impression of an "either/or" situation wherein one must either (a) have commodity, reputation, and such, or (b) have truth in life. If he were still alive, I would ask him to read my book so that I could have the benefit of his feedback (a sincere intent), and I would hope that he would also emerge from the reading motivated to rush right out to his archives and fix this "truth vs. repose" thing into "truth and repose". I think he would accept the point. Nothing more than basic TOC evaporating-cloud theory. Teach a little here. Teach a little there. Pretty soon, you've really got something (McMullen, 1997).

2   Italics and braces added. People who like and understand TOC can think of Camus' notion of "rebellion" to be akin to TOC's notions of a well-trimmed future tree and a well-executed Step 5 loop. In other words, to be a rebel in Camus' terms is to be constructive, courageous, thoughtful, and helpful in ensuring that solutions do not become counter-productive in the initiative stage and that inertia in policy or other forms of thinking doesn't just hang around to produce effects that are the exact opposite of what they were intended to accomplish. By way of contrast with Camus' vision, TOC's five-step focusing process is a little more, well, practical for use in the field. Don't get me wrong; I'm not throwing stones at Camus here. He's done the Western world — and, therefore, at least for the moment, the world — a nice service in sorting one of the safe, constructive, thoughtful, high-spirited, intelligent, and inclusive paths one can use to walk tall through life, without ignoring, getting whip-sawed by, or feeling much sting from the turbulence in the history and current state of life's ideas. Still, there's this "minimum number of simplest concepts" thing we explored during the Chapter 1 section that dealt with the dreaded "Bookstore Effect". So, take your pick. Choose either to (a) read Camus, Sartre, Nietzsche, Plato, Locke, Berkeley, Hume, Dostoevsky, Milton, Yeats, Lewis (as in C. S.), Kant, Hegel, Goethe, Tolstoy, James (as in William), Adams (as in Scott), and on and on and on and figure out what they are saying, then figure out what they are *really* saying and then somehow distill a synthesis for what to do in the field, or (b) work, at least in the practical and results-oriented portion of life, with TOC's expression of management science that offers, well, a minimum set of simplest concepts with the maximum explana-tory and predictive power. If you have the time, the energy, and the interest, then do both (a) and (b). What's the best approach? Read this book, do what it says, ramp up profes-sional performance and overall balance-in-life by doing (b), enjoy having a life, and then — in all your spare time — do as much of (a) as suits and serves you.

3   What do you mean, "Mike who?" Mike. You know, *Mike. Mike! MIKEY,* for heaven's sake! He likes it. He doesn't care about all that philosophical stuff. He likes TOC because it works! After lumbering through Confucius, Emerson, and Camus, this quotations page really needed a little lightening up, wouldn't you agree? Still, the serious moods among us are maybe not liking this Mikey thing. There may be some who have never seen the Mikey television commercial, and others who didn't like the whole Mikey thing even back then. I am asking your forbearance here because, remember, it is possible for a subject like physics to get really dull. Even the energy of irritation is better than going straight to sleep. Make you a deal. I promise I'll replace the Mikey thing in the revised edition. Thing is, I have to *get* to a revised edition in order to take it out, so — if you really dislike it — please tell everyone you know to buy and read my book so this edition will be a huge success and I can get the chance to take this Mikey thing out. Make sense?

4   Overall, K. and Papp, K., What drives your business? (Avery Dennison), in *1995 APICS Constraints Management Symposium Proceedings,* April 26–28, Phoenix, AZ, APICS, Falls Church, VA, 1995 (proceedings, page 104, APICS item #04216; Avery-Dennison audio-tape, APICS item #01511).

5   Shoemaker, L. J., It's a jungle out there ... so listen to the drumbeat!!!, in *1995 APICS Constraints Management Symposium Proceedings,* April 26–28, Phoenix, AZ, APICS, Falls Church, VA, 1995, page 123 (APICS item #04216).

6   Adelman, P. and Adelman, H., TOC case study at Hannah's Do-nut Shop, in *1995 APICS Constraints Management Symposium Proceedings,* April 26–28, Phoenix, AZ, APICS, Falls Church, VA, 1995 (proceedings, page 9, APICS item #04216; audiotape, APICS item #01505).

[7]    Noreen, E., Mackey, J. T., and Smith, D., *The Theory of Constraints and Its Implications for Management Accounting*, North River Press, Great Barrington, MA, 1995, page 102 (APICS item #03356).

[8]    Telephone conversation with Steve Smith, production superintendent, August 13, 1997.

[9]    McKinney, B. L., Klein, C. F., and Angst, D. R., Johnson Controls' experience with the Theory of Constraints, in *1996 APICS Constraints Management Symposium Proceedings*, April 17–19, Dearborn, MI, APICS, Falls Church, VA, 1996 (proceedings, page 146, APICS item #04218; Johnson Controls audiotape, APICS item #01552).

[10]   Shoemaker, L. J., It's a jungle out there ... so listen to the drumbeat!!!, in *1995 APICS Constraints Management Symposium Proceedings*, April 26–28, Phoenix, AZ, APICS, Falls Church, VA, 1995, page 123 (APICS item #04216).

[11]   Israni, A., in *1995 APICS Constraints Management Symposium Proceedings*, April 26–28, Phoenix, AZ, APICS, Falls Church, VA, 1995, page 96 (APICS item #04216).

[12]   Murphy, R. and Sines, S., Breakthrough performance in the semiconductor industry (Harris Semiconductor), in *1996 APICS Constraints Management Symposium Proceedings*, April 17–19, Dearborn, MI, APICS, Falls Church, VA, 1996, page 18 (APICS item #04218).

[13]   Shoemaker, L. J., It's a jungle out there ... so listen to the drumbeat!!!, in *1995 APICS Constraints Management Symposium Proceedings*, April 26–28, Phoenix, AZ, APICS, Falls Church, VA, 1995, page 123 (APICS item #04216).

[14]   Wilson, K., Smart business decisions at Ketema A&E, in *1996 APICS Constraints Management Symposium Proceedings*, April 17–19, Dearborn, MI, APICS, Falls Church, VA, 1996, pages 117–123 (APICS item #04218).

[15]   Schoenen, D. K. (Division Controller, Bethlehem Steel), Test driving TOC thinking processes at Bethlehem Steel, in *1997 APICS Constraints Management Symposium Proceedings*, April 17–18, Denver, CO, APICS, Falls Church, VA, 1997 (proceedings, pages 52–53, APICS item #04220; Bethlehem Steel audiotape, APICS item #01574.)

[16]   Pickels, D. and Cole, H., Reduce leadtimes and increase profits using TOC, in *1997 APICS Constraints Management Symposium Proceedings*, April 17–18, Denver, CO, APICS, Falls Church, VA, 1997 (proceedings, pages 22–23, APICS item #04220; audiotape, APICS item #01567).

[17]   Gallagher, N. A. (General Manager), TOC implementation at ITT Night Vision, in *1997 APICS Constraints Management Symposium Proceedings*, April 17–18, Denver, CO, APICS, Falls Church, VA, 1997 (proceedings, pages 10–11, APICS item #04220; ITT audiotape, APICS item #01565).

[18]   Pirasteh, R. M. (Plant Manager) and Camp, G. B. (General Manager), Integration of synchronous manufacturing (SM) and application of the Theory of Constraints (TOC) at Dresser Industries, in *1997 APICS Constraints Management Symposium Proceedings*, April 17–18, Denver, CO, APICS, Falls Church, VA, 1997 (proceedings, pages 5–8, APICS item #04220; Dresser Industries audiotape, APICS item #01563).

[19]   Wilson, K., Smart business decisions at Ketema A&E, in *1996 APICS Constraints Management Symposium Proceedings*, April 17–19, Dearborn, MI, APICS, Falls Church, VA, 1996 (proceedings, page 122, APICS item #04218; Ketema A&E audiotape, APICS item #01550).

[20]   Shoemaker, L. J., It's a jungle out there ... so listen to the drumbeat!!!, in *1995 APICS Constraints Management Symposium Proceedings*, April 26–28, Phoenix, AZ, APICS, Falls Church, VA, 1995, page 123 (APICS item #04216).

[21]   Moon, S. A., TOC at Parr Instruments: a view from the inside, in *1996 APICS Constraints Management Symposium Proceedings*, April 17–19, Dearborn, MI, APICS, Falls Church, VA, 1996 (proceedings, page 63, APICS item #04218; Parr Instruments audiotape, APICS

item #01539). Stacey Moon, a brilliant woman who has shaped a fine career in production management from beginnings in entry-level clerical assignments, learned TOC from one of the best. The man she describes as her boss and mentor is one of the pioneers from the early Valmont Industries TOC projects, Jeff Wood.

22  Wayman, W., As excess capacity is exposed, how do you increase productivity?, in *1996 APICS Constraints Management Symposium Proceedings*, April 17–19, Dearborn, MI, APICS, Falls Church, VA, 1996 (proceedings, page 154, APICS item #04218; Morton Automotive audiotape, APICS item #01554)

23  Murphy, R. and Sines, S., in *1996 APICS Constraints Management Symposium Proceedings*, April 17–19, Dearborn, MI, APICS, Falls Church, VA, 1996 (proceedings, page 18, APICS item #04218; Harris Semiconductor audiotape, APICS item #01536).

24  Noreen, E., Mackey, J.T., and Smith, D., *The Theory of Constraints and Its Implications for Management Accounting*, North River Press, Great Barrington, MA, 1995, page 103 (APICS item #03356).

25  Shoemaker, L.J., It's a jungle out there ... so listen to the drumbeat!!!, in *1995 APICS Constraints Management Symposium Proceedings*, April 26–28, Phoenix, AZ, APICS, Falls Church, VA, 1995, page 123 (APICS item #04216).

26  Moon, S.A., TOC at Parr Instruments: a view from the inside, in *1996 APICS Constraints Management Symposium Proceedings*, April 17–19, Dearborn, MI, APICS, Falls Church, VA, 1996 (proceedings, pages 57, 63, 64, APICS item #04218; Parr Instruments audiotape, APICS item #01539).

27  Maybe Mike worked for one of the companies that had this happen. That's it this time. Really.

28  This case information is disguised at the request of the company. The reasons? The usual. The division executive supported the plant's use of TOC but hadn't had a chance to get an understanding of TOC. Above her, the group executives were aware of TOC, since one of the major groups was instituting TOC measurements, but they had not had a chance to be briefed fully on TOC, either. (This book will help a very great deal with that sort of problem.) Of the people in the company who did understand TOC, many learned much of what they knew about it from competitors' presentations at APICS Constraints Management symposia. As a result, they believed they had in TOC a time-based competitive advantage (sort of a time delay advantage) in instituting TOC and did not want to tip their hands to their other competitors. Finally, while the financial and operational performance of the plant had, in fact, reached such a low point that executives were at least beginning to wonder about the future of the factory and its product lines, they had not felt the situation had progressed far enough to make formal or even informal communications to the employees or the local community in question. In other words, there were several legitimate reasons the company asked that this data become unrecognizable. Still, hats off and many thanks to the companies who have been willing to put their experience more fully into the public domain so that the rest of the world can learn from it. There is likely a speed advantage, which is the other side of the coin from the time-delay advantage that exists and increases — as a "chicken or egg" type of thing — when a company has the confidence and positioning to be able to put information like this into the public domain. Interesting set of choices. I am not saying either choice is wrong; however, I *am* saying that the companies and individuals who have chosen to present and share at APICS Constraints Management symposia have made a huge contribution to the world's understanding and ability to use TOC. Thank you.

29  Millstein, I.M., Distinguishing "ownership" and "control" in the 1990s, in *Institutional Shareholder Services (ISS) Conference Proceedings*, February 25, 1994; reprinted as Appendix

7 in Monks, R. A. G. and Minow, N., *Corporate Governance*, Blackwell Publishers, Cambridge, MA, 1995.

30  Reichheld, F. F., *The Loyalty Effect: The Hidden Force Behind Growth, Profits, and Lasting Value*, Harvard Business School Press, Boston, MA, 1996. The "The Right Employees" chapter means what it says. Along with pointing out that cultivating the loyalty of some employees has economic advantages, the author notes that reducing defection rates of other employees does not create value (see page 102): "The old promise of lifetime employment attracted people who, in the new competitive environment, no longer earn their salaries. Cutting their defection rates will not create value." I think Reichheld's comments are correct for the 1990s period in which it was written. I think they will not be correct for the economic future ahead. Once the people who were trained by employers and society to think and behave in ways that matched the promise of lifetime employment have been equipped with better ways to think about themselves and their work, the economic and institutional winners will be those who gain the economic benefits — via the same seven-factor, cause-and-effect economic model discussed in the Bain & Company study results on pages 100 to 102 of *The Loyalty Effect* — from employees and other key stakeholders enjoying the realities of open-ended, if not lifetime, employment. That this seems not possible at the moment in our western society and economy does not make it any less the inevitable right answer. Loyalty will again go both ways. It will apply to the full spectrum of people in the workforce and will be the right thing, not only in terms of efficiency and effectiveness, but also in every other way that matters.

31  The days-long conference in April 1990 was called *Data Is Not Information*. However, instead of simply reflecting the results of the field research and software prototype development that were underway at that time (leading to *The Haystack Syndrome*), the event became a classic instance of Dr. Goldratt taking it from the top, verbalizing his intuition as it had been shaped and synthesized by all the activities and thinking of the recent years and re-formulating on the spot, in front a few hundred managers — using the usual two overhead projectors, blank slides, and multi-colored marker pens — a latest positioning statement for what TOC, the emerging new management science, had by then become. If I am not mistaken, there were videocameras there, so someone somewhere most likely has the tapes.

32  This is only one of the more obvious ways in which every individual does, can, and should make a difference.

33  That is a quote from Dr. Goldratt.

34  Mueller, R. K., *Anchoring Points for Corporate Directors: Obeying the Unenforceable*, Quorum Books, London, 1996.

35  I can hear the hoots already. But don't be so hasty. Think about it. The only way to have something work is for most people to do it because it's right. Relying on rewards and punishments may have its place, but it is just not in the same league with having people doing something because they have thought about it and believe it is right.

36  The article appeared originally in the March/April 1991 edition of the *Harvard Business Review*. It is also reprinted on pages 109 to 125 in Reichheld, F. F. (Ed.), *The Quest for Loyalty: Creating Value Through Partnership*, Harvard Business School Press, Boston, 1990–1996.

37  I like it a lot that the TOC TVA financial management system that I am helping to recommend to industry looks a lot like what Peter Drucker is recommending in *Harvard Business Review* articles to boards of directors, institutional investors, audit firms, and, indirectly, government policy makers. I make the tone of this tome sometimes light and lively to keep us all awake and to make the ideas more likely to be used effectively by a lot

of different kinds of people. What that does *not* show is that I have spent, and continue to spend, a lot of time thinking about the potential — even the remote potential — effects of TOC becoming a global baseline management best practice, in a lot of different areas of life. Why? Well, one would not like to have one's gifts for understanding, applying, and helping to evolve something *and* one's gifts for helping that something become a global best practice turn out to have been applied to something that produced something other than constructive changes. That a thoughtful observer such as Drucker is saying roughly the same thing in different words is helpful in this regard.

38   Incidentally, I did not get into it in Chapter 1, but TVA also occurs when as asset is sold as Sales minus Total Cost of Asset (the capital investment portion of the "Inventory" portion of TOC TVA-I-OE scorecards). I am just mentioning that in case someone is reading Drucker's quote to include, within the notion of "wealth producing capacity", the creation of assets that generate profits upon sale of the assets or of the entire going concern, but without ever being realized as the type of inflowing TVA that comes from product or service sales. I can't cover it all in this book, but, yes, I know some of you are out there visiting all my accounting comments with your capacity for steely-eyed precision. Stay tuned.

39   My family is a union family. Over the years, life has led me quite a little distance from the union lifestyle — that is if you leave out family reunions. Still, no amount of nonsense emitted from urban and other professionals can remove the knowledge that there can be no useful stereotype of the working man or woman. There is no legitimate single view of what a union person is or thinks. In a word, they are all people. Which means they have something in common with managers, executives, Wall Street pension fund managers, doctors, lawyers, Native American chiefs, butchers, bakers, and tool-and-die-makers. Not to mention all the other people in life. *All* people. The practical idealist segment of every one of these groups of people, in every one of these walks of life, should be able to see the common ground in TOC and work together to make TOC one of this generation's contribution to improved standard practice.

40   Stevenson, H. H. and Moldoveanu, M. C., The power of predictability, *Harvard Business Review*, July-August, 1995; reprinted in Reichheld, F. F. (Ed.), *The Quest for Loyalty: Creating Value Through Partnership*, Harvard Business School Press, Boston, 1990–1996, pages 67–72.

41   TOC logic trees are not the first instances of visual tools for having employees in departments understand the impacts of their actions on overall outcomes. There is the famous example of Jim Treybig, founder of Tandem Computer, creating a huge diagram that is said to have helped explain how each employee mattered to the overall company performance. Treybig was the person in the company who had the facts and intuition about how it all worked, in terms of both tangible and intangible factors and in terms of factors affecting both current and future performance. He verbalized that intuition, empowered his people, was most likely further empowered himself by the very experience of converting his view from implicit to explicit (verbalizing intuition does have that effect, by the way, a fact not to be ignored), and had a pretty darn good visual aid that was valid for long enough to move the company forward to some new and healthy state of operations. So TOC's logic tree thinking processes are not historically the first of their class of tool, but they are the best and most practical means for more and more types of people to verbalize intuition and other facts about complex systems. After all, although every company needs for its people to understand the critical relationships among departments and overall company goals, we can't all be Jimmy Treybig.

[42]    Frontispiece praise printed inside the front cover of Goldratt's *Theory of Constraints: A Systems Approach to Continuous Improvement*, the valuable source book on the TOC logic-tree thinking processes published in 1996 by H. William Dettmer (APICS item #03516).

[43]    This is a comment made on the Internet by Larry Leach, Consultant, Quality Systems, Idaho Falls, ID.

[44]    McKinney, B. L., Klein, C. F., and Angst, D. R., Johnson Controls' experience with the Theory of Constraints, in *1996 APICS Constraints Management Symposium Proceedings*, April 17–19, Dearborn, MI, APICS, Falls Church, VA, 1996 (proceedings, page 146, APICS item #04218; Johnson Controls audiotape, APICS item #01552).

[45]    If you were a conscientious objector in the wars to create the New Age or if you never noticed they happened, then you should skip this next part. You won't understand it, anyway. On the other hand, if you either loved or hated the New Age thing, read on. One warning, though: This footnote is the McMullen Associates' application for entry into the *Guinness Book of Records*. Category? World's longest footnote, naturally.

    *Didn't Like New Age?* If you really did not like the New Age thing, then your best move is to get all the people around you who did walk those paths to read this and other TOC books. You'll be glad you did. So will they, but don't let that stop you. There are many people out there who have been deeply influenced by the New Age spiritual themes. For them, there is no returning to what they view as more limited forms of experience. Plus, many honestly believe that the heart of the experience they feel strongly about cannot be found or cannot be found without side effects, within the contexts of the traditions they learned in their youth. Which means, short of solving the problem through civil strife (which I definitely do not think is a good idea) and short of throwing the cultural baby out with the cultural bathwater (which I also do not recommend), there is a real need for a constructive and effective way to sort out the timeless eternal values that are both common and important to the fundamental motivations of both the newer and the traditional modes of experience. Enter TOC and the logic trees, or something like it, which can be used over a period of a generation or three, to get under the assumptions of both the New Age and the Old Age (?) groups, to increase the range or depth of spirituality of some (not all) of the traditional ways, and to increase the practicality, effectiveness, and comprehensiveness (as, in some cases, including the facts and ramifications that other people are out there/here, too) of some (not all) of the newer ways. Generation or three? Sure. Some things take time, especially when the types of steps typically taken can't — as they say in Maine — get you there from here. (What's that? Get the pronunciation right on the Maine thing? Try this: cahn't get yuh theyah from hee-uh. Better? Bettah?) As best I can tell, the world is only as "nice" and as "fair" as people cause it to be. People will always want to cause it to be as nice and as fair as they think it can reasonably be. If they think things could be more nice and more fair but traditional things are blocking getting there, they will view the traditions as "unfair". Whether they are right or wrong about that, they will move down paths outside the tradition. When they go there, they find a lot of things. If some of the things are tools (not talking about TOC tools here) that make them more effective, and feel more well, and feel more strong, and feel more happy; this will make them wonder why the tradition of their youth somehow had inadvertently placed obstacles to the discovery and use of those pretty much natural and universal tools. Maybe — and maybe even for what were perceived at the time as some good, practical, and necessary historical reasons — the obstacles were not placed there inadvertently, but rather on purpose. Dostoevsky's Ivan, in the Grand Inquisitor scene of *Brothers Karamazov*, seems to be telling us *he* thinks so. But, in any event, that was then, and now is now. Keep in mind that the tools some find outside the traditions are tools that also increase their personal effectiveness in everything, including sometimes in efforts to push back against the traditions. Think about that. Maybe carefully calibrated adjustments to some of the traditions that may amount only to removing a slice of the over-conservatism — which was maybe placed to prevent something bad, but was either more than necessary to prevent the bad or was not the only way to prevent the bad, or both — can be made. If the adjustments are well crafted, maybe they fit within some portion of the traditions anyway and therefore aren't really adjustments anyway. Maybe they remove some of the objections that drove people outside the traditions anyway. Maybe they restore to the traditions some aspects of individual and group effectiveness that, over time, would otherwise have amounted to unnecessary competitive disadvantage and that would have incrementally increased questions both of a tradition's effectiveness and longevity. Maybe the adjustments could also move the traditions forward in the area of being both effective and being perceived as being effective in the areas of creating the appropriate degrees of "nice" and "fair" and "just" in the world. Maybe, just maybe, every tradition should make sure it understands what each of the many and different so-called New Age phenomena are telling it about itself, about its place in the world, about its relative vitality and effectiveness, and about the likelihood of its being viewed as effective in increasing the range and depth of those "nice" and "fair" and "just" characteristics in the world, characteristics that virtually everyone in the world care about. That's, of course, unless you really don't want to. Decisions, decisions.

*Liked New Age?* Meanwhile, and on the other hand, let's say you did like some of the New Age phenomena. If you got anywhere near any or all of the thousands of different spiritual emergence, enlightenment, self-actualization, life-as-illusion, east-meets-west, personal transformation, social transformation, Age of Aquarius, new physics, spiritual healing, alternative medicine, neuro-linguistic programming, cybernetics, gestalt, inner game of tennis, inner game of skiing, inner game of inner games — anything like that — then here is why you are going to like the TOC logic-tree thinking processes. Maybe you read Fritjof Capra and learned that quantum theory is a lot like one of the classical mystical experiences. Maybe some of those quanta even helped you get there yourself. Fine, once you're back *here* (sort of) from having arrived *there*, you still need to know how you're going to convert the new you into practical actions in the world. TOC and logic trees are lovely for that. Enjoy. While we're doing things quanta, if you read Richard Bach's latest and discovered that the most subtle insights about the view of life and life-as-illusion can be captured in only a few dozen words, in terms akin to those of quantum physics, then congratulations. Ditto, however, from the above. What to do now? Gotta pass the time somehow. Enlightenment is one thing. Couch potato is quite another. So break out the TOC TP. If you liked visiting with Ian Malcolm, Michael Crichton's ever-present and peripatetic chaos theoretician in *Jurassic Park* and *The Lost World*, you can join the TOC consulting firm, Chesapeake Consulting, in blending TOC with chaos theory to great effect in industry and other areas of life. Why should all those darling butterfly wings in China have all the fun? Let's keep moving here. If you were transformed by Werner Erhard's est or if James Redfield's *Celestine Prophecy* is your favorite flavor of ordering data and making predictions, then the TOC thinking processes will help to shape your incoming patterns of experience and meaningful coincidence. If you were made clear by Hubbard's *Dianetics*, you'll want to start right away to use the TOC thinking processes to construct hay from your greater freedom from involuntary behaviors, and your greater clarity of mind and experience. Better yet, if you have mastered Hubbard's material, do what Herbert Benson did for transcendental meditation and extract the cause-and-effect relationships from within all the difficult style issues and make any powerful thinking solutions to health and mental health a lot more understandable and palatable as practical alternatives to your least favorite psychotropic drugs, if they are. If you are the *Jonathan Livingston Seagull* style of Richard Bach fan, TOC thinking processes will help you select and fine-tune your selected type of flight. If you are of the later and riskier Richard Bach *Illusions* camp and style, then the TOC thinking processes will help you nail down exactly what you mean by your preferred blue feather and — for complex feathers — break down the gargantuan task of getting projects done. As a side benefit, TOC will also assist with evaporating all manner of clouds. Finally, if you're of the even more lively *Running From Safety* Bach era, TOC thinking processes make it much easier to distinguish between those things to which you're planning to provide consent, and things to which you are planning to deny consent, no small contribution. If you liked Rupert Sheldrake, your work with TOC tools will help to influence which things get coincidentally discovered in how many different places on Earth, and when. If you liked Maharishi Mahesh Yogi's transcendental meditation and maybe even attended his university, TOC logic trees can fine-tune the objectives sought as individual and cooperative efforts arise to make use of the greater clarity of mind and the deepened and less cluttered sense of self. Ditto for the Four Noble Truths, the Eight-Fold Noble Path, and things Tantric, and Ch'an, and Zen. Shifting gears to a different and trickier domain, maybe you never told anyone, but maybe you listened to a lot of Moody Blues tunes and liked the sounds a lot. What was it about the feelings they produced? Maybe you also noticed some of the words. Watch your step, but maybe you understood them. Get ready to duck, but maybe you *really* understood the words. If so, you can maybe participate in a timely TOC negative branch process in which it, in some generation, becomes clear which patterns of thinking will lead to the same peak, but more importantly to the same resultant everyday, average experiences (desirable effects) as the route or routes you took. Ditto for any and all elected or almost-appointed public officials and everybody else who either inhaled things or not, played saxophone or not, or both. If you've followed the encyclopedic Marilyn Ferguson's reports over the years (I did for a while), TOC thinking processes will provide an orderly way of sorting out which parts of the *Aquarian Conspiracy* are left to work on and which are already done deals. Good news and bad news here, for these folks. The good news first. One can relax about whether the New Age is going to succeed in getting here, despite all the best efforts of all those Old Age reactionaries (whoever they were). Now for the bad news — now there's a need to get on with the work of sorting out how to have a sensible life society in the midst of the new spiritual, ethical, and temporal chaos. Okay, it's wide open now. The state of the western world now gives a whole new meaning to the American phrase, "The Wild West". The challenge now is to see how much of the New Age material has to be downplayed, or even set aside, in order to know how to answer the question, "What are we going to teach our children and grandchildren?" You may be able to *not* think about this, but not everyone will be able to or will want to *not* think about it. If you are willing to think about it, a new problem pops up. Many non-New Age, wanna-be statesmen and executives have suddenly found themselves in charge, and discovered that — in their single-minded focus on the issue they used to get into the saddle — they had never learned the issues and answers involved in governing. The New Age finds itself now in the saddle to some extent. Do they know what to do with it, and how? As Daniel Quinn's *Ishmael* asks, What world will the New Age now create? Knowing of a lot of the useful, but sometimes narrow or sometimes uneven New Age views and tools, it is possible that many of the people who believe in them with intensity haven't thought about the combinations of things that need to be in place in order for a world to make sense for themselves, for everyone else, and for everybody's children and grandchildren. Use TOC to discern the helpful administrative realities that match the stages of evolution of the human family's composition. I'm skipping a lot of themes here, but that's enough. Let's move on from the New Age to other places where TOC logic trees can serve usefully.

*TOC logic trees are useful virtually everywhere.* Shifting gears to traditions, if Judeo-Christian faith is your way of thinking about life, then the TOC thinking processes will help you focus on (figuratively) which mountain you are choosing to move, which sea you are likely to part, which manna should arrive from heaven, which water will turn to wine, and so forth. Don't get mad. Think about this. If you've seen the world the way Swedenborg did, or you've experienced that shift and surge in energy and life that happens from encounters with Mary Baker Eddy's *Science and Health* — complete with all the Christian Scientists' blue crayons, metal page turners, and side-by-side books and bibles (been there; done that; yes, all the news items aside about taking the "no medicine" strategy to extremes that have tragic consequences, there is very definitely something powerful going on there) — then use the TOC thinking processes to sort out what to work on and how. While you're at it, you would do well to do here what Herbert Benson did for transcendental meditation and better clarify the cause-and-effect that lives within the Science, if what is working there is intended to have an increasing positive impact in the world. I noticed in *Retrospection and Introspection* that the admonition exists not to tinker with the strategy of exposition contained in the latest revision of *Science and Health*. However, as mentioned in the Chapter 1 section on Goldratt's baseline TOC, can any one person — even Mrs. E — get it so right, in some single way, that will be the only right way for everyone, in all circumstances, for all times? If you sympathized with the *Mahabarata's Bhagavadgita*'s Prince Arjuna's predicament on the battlefield, receiving the celestial advice of charioteer Lord Krishna, who tells him he should just carry out his role, and became concerned, then here is an idea. You can use the TOC logic trees to help you to identify and get into the role in life in which you'd be comfortable taking the charioteer's advice yourself. Speaking of sympathizing, I happened to be walking down a sidewalk at the Los Angeles International Airport — a young Navy officer in search of the big bike box with my 10-speed in it — at exactly the moment a scene for the movie *Airplane* was being shot and re-shot and re-shot. It was the one in which Robert Stack, as an airplane pilot trying to get through the airport to his plane, performs remarkable martial arts moves to stave off the people accosting him with flowers and long flowing robes. Amazing. The poor guy in the robes took a beating; he should have used TOC to sort out other strategies for maintaining steady or growing TVA under all likely combinations of economic and missionary circumstances. Just kidding, but I *did* see the Robert Stack shoot. It was great. Back to the story.

*People of power:* If you are a man or person "of power" due to a return from Casteneda's cactus, Nietzsche's *Zarathustra*, Sade's *Justine*, Emerson's *Self-Reliance*, Silva's nocturnal interpersonal messaging system (or Jung's, for that matter), tai chi's jing, Tony Robbins' burning walkways, Darth Vader's or Luke Skywalker's or Obe Wan Kenobe's force, Siddhartha's samana-depressor, various pincushion or spell technologies, red cell, green cell, aikido, Segall, samurai, or even non-oriental tough guy stuff — that's all to the good, maybe, but there is still the question of what to change and what to change to through the use of all that nice new and "secret" power (there's so much of this going around now, in so many forms, that it's not much of a "secret" anymore, y'know?), and how to work with others to do more together than even you can do by yourself. Naturally, TOC logic trees belong here, too, to guide, focus, and sequence the consensus and efforts. Which means we need a new book? *TP for Tough Guys? (and Gals)?* If you came away savvy from Maccoby's *Gamesman*, really effective from Kotter's *Power*, sly from Machiavelli's *Prince* (or the recent update, *The Princess*), there are still the questions of what to change and what to change to. Answer? You guessed it. TOC thinking processes (there's a pattern developing here).

Did I forget to mention anybody? I know what you're thinking: He couldn't possibly have missed anybody. But then, if you liked the course in miracles, the way seth spoke, the candles, the crystals, the 1-800 and 1-900 for psychics, astrology, ouija board, or tarot, there are still those pesky questions about what to do with the extra sources of advice. One last person: James Dean, the rebel without a cause. If he had just verbalized his intuition, he could have made explicit what was bugging him. He could have identified a cause, a core problem even. It may even have been a fairly deep cultural cause, maybe even a cause so subtle that it causes an — I don't know — an existential malaise (say that three times fast). It may have been a cause so tough it would be tough to tackle it, especially if nobody wanted to acknowledge that it's a problem. He could have moved briskly from current tree, to cloud, to future tree, all the way to a pre-requisites logic tree to tackle that problem one bite at a time. He could have become a "rebel *with* a cause" and made another movie. Okay, go ahead and laugh. But if you think James Dean is gone, look around and think again. There are a lot of people who really believe that the problems in the "system" are too big to deal with, and, therefore, they don't. That, by itself, is a problem. But that's kid stuff compared to the problem of what people do with their spare time, to pass the time, when they see things (incorrectly) that way.

Here's one for the Ayn Rand community. You have become remarkably — no astonishingly, no astronomically — competent in the image of a Dagny Taggart, Hank Reardon, Francisco D'Anconia, John Galt, and all the rest of the wizards without whom the world supposedly just won't run. (Obviously, Gary, they never read your elegant poem about the hand removed from the bucket of water.) Nevermind the *Fountainhead* people for now; let's just consider the *Atlas Shrugged* crowd. Now what? Just because the Jim Taggart's of the world are pulling the strings with sneaky moves, you are actually going to let Atlas shrug? Hey, don't let the politics get to you like that! Sure, it's awful and it's constant, but so is flossing, and you don't just up and quit humanity over that. How about just demoting Jimmy Taggart's importance? Place him and his hijinks (and the real equivalents in your own life) into an entity or two on a TOC current-reality logic tree, get to a future-tree vision of what you are going to work on with all that hard-won competence, measure your progess with dollars if you must, then work out a transition path that works around Dagny's slippery lousy political snake of a gawdawful brother. After all, that little place you and all your fellow mega-competents are all going to retire to is going to get a little dull. Do you really think all you movers and shakers and world-holder-uppers are going to have a good time out there with virtually nothing to work on? Hey, there's plenty to work on over here in this world. It is reasonable to assume that the good Maker, whatever you call him or her (actually, you don't call him or her anything, do you?) ... it's reasonable to assume that you are on the Earth for a reason, and it's NOT to take your football and go home, only to return from time to time for a few unscheduled radio broadcasts that are going to go way over most everyone's head anyway, especially if you keep letting them go on for so long. (Kind of like this footnote.) Keep in mind, when you declare, "This is John Galt," you had better know that *that* only gets your foot in the door. It doesn't necessarily win you a long span of rapt attention. More than all that, though, the TOC logic trees are a vehicle for synthesizing the objective and subjective and, therefore, for deploying the combination toward objectives of importance to you. That's not something to just shrug off.

Okay, that's enough of this. At last! We didn't get to all of the points, but we got to a few. The main point is, for better or for worse (and it is a very important open question as to whether it *is* better or worse), that the battle for diverse views and — much more than that — the battle for a more diverse experience has been won. Now the question is, "What are this generation and the coming generations going to do with all this newfound freedom to be?" Shall we fight over it? If so, the title of Laing's book, *The Politics of Experience*, takes on new meaning and relevance. Or shall we discover, as each person better and more deeply understands his or her own inherited or chosen view, that he and she also better and more deeply understand and more easily accept the other's view? How will all these newer (or, some will insist, newly resurgent) spiritual insights, experiences, and lifestyles be blended somehow with what have been for a long time the major traditions, to evolve into a "smaller" world that people (a) believe they understand, and (b) believe makes sense? By making common sense a common practice? Is there another way? If not, what approach to managing, in industry and anywhere else in life, seems, at least at the moment, to have the highest likelihood of being able to make common sense a common practice? Any guesses as to which one *I* think it is?

46   Wheatley, M. J., *Leadership and the New Science*, Berrett-Koehler, San Francisco, CA, 1992.

47   For those who like to mix their chaos theory with their TOC, James Gleick's book is an excellent source for the chaos portion of the brew: *Chaos: Making a New Science*, Penguin Books, New York, 1987. A few TOC logic trees would be a nice addition to that book's next edition.

48   Block, P., *Stewardship: Choosing Service Over Self-Interest*, Berrett-Koehler, San Francisco, 1993.

49   Covington, J., Leadership and the next generation of strategic planning, in *1996 APICS Constraints Management Symposium Proceedings*, April 17–19, Dearborn, MI, APICS, Falls Church, VA, 1996 (proceedings, pages 66–77, APICS item #04218; audiotape, APICS item #01541).

50   Israni, A., Applications of the Theory of Constraints in the semiconductor industry (National Semiconductor), in *1995 APICS Constraints Management Symposium Proceedings*, April 26–28, Phoenix, AZ, APICS, Falls Church, VA, 1995 (proceedings, APICS item #04216; National Semiconductor audiotape, APICS item #01509).

51   Frontispiece praise from Bill Dettmer's 1997 thinking process book (APICS item #03516).

52   Cox, J., Mordoch, A., Spodeck, G., and Kingman, O., TOC panel discussion on the use of TOC logic tree thinking processes for business field research, in *1996 APICS Constraints Management Symposium Proceedings*, April 17–19, Dearborn, MI, APICS, Falls Church, VA, 1996 (proceedings, pages 78–82, APICS item #04218; audiotape, APICS item #01542).

53   Hinneburg, P. A., Lynch, W., and Black, J., Lean logistics in the U.S. Air Force Logistics Command, in *1996 APICS Constraints Management Symposium Proceedings*, April 17–19, Dearborn, MI, APICS, Falls Church, VA, 1996 (proceedings, pages 89–97, APICS item #04218; audiotape, APICS item #01548). Update: General Hinneburg, the pioneering Air Force officer, is now employed at the Boeing Corporation on the west coast.

# Notes for Chapter 4

1    An American phrase for "directly from the person involved".

2    The other primary force in this implementation was a handsome TOC consultant and educator named Tom McMullen.

3    Bob Vornlocker was not allowed to be audiotaped or to submit a paper (we allowed that in a few cases in the first APICS event); he is mentioned here to acknowledge his significant early contributions to the TOC manufacturing systems industry, and to the first APICS TOC Symposium. Besides, he's a great guy. Hmmm ... While I'm at it, I don't think Avraham Mordoch is mentioned in the book now that I decided to trim Chapter 2 in favor of expanding Chapters 1 and 6. Sorry, Avraham, it was either Ethan Hunt or Avraham Mordoch. I knew you would understand. Anyway, Avraham was the Fearless Leader leading the charge on *The Haystack Syndrome* systems development project. Tracey Burton Houle, Andy Macri, Rob Newbold, Mike Bannis, Ronni Halavi, Eli Schragenheim, Bob Fox, Dale Houle, Harriet Harris, geez, the gal from the Connecticut consumer products company (oh dear, they say memory is the second thing to go) ... gotta keep moving ... these people, plus Dave Ward, Frank Ward, Morry Riehl, Tom, Tom Spangrud, Jack Buntin, Ed, Glen, Tom Madden, Terry Frick, Gary Smith, Lynn White, Gene Makl, Jerry, Paul, and many more also made important early contributions to *The Haystack Syndrome* class of TOC systems.

4    The author was the conference chairman and one of 16 presenters at this event.

5    The magazine was the *International Journal of Schedule-Based Decision-Support* (*IJSBDS*). The title reflected its purpose in introducing TOC at its higher levels, including strategic planning, in order to show how the lower levels of drum-buffer-rope and buffer management fit into the overall picture of a TOC management system implementation and superior ROI. Alas, the publication was set aside due to large events and new responsibilities which introduced themselves into my personal life during the 1992 and early 1993 time frame.

6    Plantz, B., Custom kitchens in 10 days: Kent Moore Cabinets uses synchronous manufacturing, *Furniture Design & Manufacturing*, October, 1991; also provided in the proceedings of the Institute for International Research (IIR) TOC Conference held January 23–24, 1992.

7    The Institute of Management Accountants (IMA) Information Center may be reached at (800) 638-4427. In 1997, the rate for IMA members was $15 for 2 weeks' rental.

8    McMullen, Jr., T. B., A Tale of Two Best Practices, presented at the 1994 International Society for Systems Improvement (ISSI) Conference, June 1994, Fort Walton Beach, FL. The author was the TOC educator, consultant, systems integrator, change agent, gadfly, and coach for this very rich and interesting project.

9    Hamilton, D., Elwood City Forge: Using TOC and TP To Improve Customer Service and Quality, presented at the 1996 North American Goldratt Institute Jonah Upgrade Conference, March 18–21, 1996, Washington, D.C.

10   Gadbois, M., Shipments flow smoothly at Flow International, *IIE Solutions*, March, 40–43, 1996.

11   Miles, R. H., *Corporate Comeback: The Story of Renewal and Transformation at National Semiconductor*, Jossey-Bass, San Francisco, CA, 1997. Minor mention of TOC (and only in context of cycle time reduction) on pages 175–176.

12   APICS Atlantic Coast Symposium (ACS), March 1996.

13 Koziol, D. S., How the constraint theory improved a job-shop operation, *Management Accounting*, May, 44–79, 1988.

14 Henricks, M., Full steam ahead: is something holding your company back? Constraint management helps you break loose, *Entrepreneur*, July, 75–77, 1996.

15 Mealy, D. and Sundararajan, S., Strategically integrating the Theory of Constraints, just-in-time, Total Quality Management, and participative management for competitive advantage, in *1995 APICS International Conference Proceedings*, October 22–27, Orlando, FL, APICS, Falls Church, VA, 1995 (proceedings, APICS item #04019; audiotape of session is D-15 at (800) 776-5454).

16 Newsletter of the Los Angeles Chapter of APICS, September 1993.

# Notes for Chapter 5

1 It is Tolstoy's little ditty, *War and Peace*, that makes the useful point that the combined wisdom and will of large numbers of people creates the leaders and events we see. It's a good thing that a mechanism seems to exist for our aggregate will and intention to influence things. Why? Because I have known good and smart people at the top and bottom of the world who all still manage to miss really important issues by a country mile and for very long periods of time. Any tool kit (such as TOC logic trees) that helps people to know and act on what is in their hearts or to know and act on what eventually comes into their hearts as they pay attention to finding out what is in their hearts (whew!) contributes to creating good things in life. Such tools help each of us to overcome and balance out the effects of the other guy's blind spot. Fortunately, *I* don't have any blind spots. Or, if I do, only my readers really know of them. That means that the more people who read my book, the more people who are helping to cancel out the effects of my blind spots. So, will you please ask all your friends to buy and read this book right away?

2 Ayn Rand attributes it to theologian Reinhold Niebuhr in her 1973 essay, "The Metaphysical Versus the Man-Made", in *Philosophy: Who Needs It* (Penguin Books, New York, 1982). While noting the irony that she is approvingly quoting someone with whom she disagrees on every fundamental issue, she does not provide a footnote to trace the source. Her version of the quote — "God grant me the serenity to accept things I cannot change, courage to change things I can, and wisdom to know the difference" — is pretty close to my version of the popular saying.

3 This was over half a decade before I would hear about Eli Goldratt and TOC. It was a full decade prior to the emergence of the TOC thinking processes. The point is that many people, in many walks of life, have been using breakthough thinking processes that are very similar to the TOC version. The difference is, with the process now verbalized or made explicit, the activity is moved from the remarkable "art" of some specific individual to something more people can do and something more people can do together. This is a huge difference with huge ramifications, both for the intuitively bright who have not considered that everyone could do what they can do and for those who will become more competitive problem-solvers!

4 Benson, H., *The Relaxation Response*, Avon Books, New York, 1975.

5 The method was transcendental meditation, or TM. TM became well-known in the U.S. because, among other reasons, it was popularized by Maharishi Mahesh Yogi and was acknowledged by computer software industry entrepreneur Mitchell Kapor (of Lotus 1-

2-3 spreadsheet fame). The method involved reciting a phrase (mantra) for 20 minutes twice per day, which is about as simple and natural as medicine gets. The benefits are documented in Benson (1975) cited above.

[6]  Benson, H., *Timeless Healing: The Power and Biology of Belief,* Scribner & Sons, New York, 1996.

[7]  Benson, H., *Timeless Healing: The Power and Biology of Belief,* Scribner & Sons, New York, 1996, page 23.

[8]  "Driven to distraction" is a phrase associated with some of the attention deficit disorder (ADD) phenomena. If Ethan Hunt could get out of the mess he had in Chapter 1, maybe people with ADD and more serious thinking-related problems with skills and mental health can use TOC logic trees to improve situations that are contributing to their problems. Maybe for themselves. Maybe with a little help from their friends. *Knots* is the title of one of psychiatrist R. D. Laing's somewhat controversial books. *Knots* is one of many brilliant sources around that contain good partial analyses and good partial solutions, but only partial; they have had some poor history, well-deserved or undeserved poor PR, and poor packaging and represent general-purpose missed opportunity.

[9]  This sort of thing needs a foundation. Something like a Barbara Y. McMullen Foundation for Individual and Family Self-Help.

[10]  Couldn't resist the "worldwide forest" thing. I mean, if you were writing this, would you let that kind of opportunity pass?

[11]  This topic is covered in McMullen, Jr., T.B., The systems industry is giving MRP (and ERP) implementers a choice: traditional or TOC systems, in *1996 APICS Constraints Management Symposium Proceedings,* April 17–19, Dearborn, MI, APICS, Falls Church, VA, 1996 (APICS item #04218). Also reprinted in *Selected Readings in Constraints Management* (APICS item #05021).

[12]  Kathy Suerken's letter of June 30, 1992, to Dr. Eli Goldratt got the ball rolling.

[13]  For example, when in the planning of the 1996 APICS Constraints Management Symposium, Dr. Goldratt declined yet another honorarium for providing a half-day keynote program, we countered with a donation to TOC for Education, which he accepted.

[14]  Most of what follows in the TOC for Education section is drawn from telephone, fax, and e-mail communications with Kathy Suerken, December 5–7, 1996.

# Notes for Chapter 6

[1]  Not only should Yoda be credited with this one, but also the executive management of Advanced Micro Devices (AMD). I had the good fortune to work for that giant semiconductor manufacturing company once. The AMD managers may not have coined this phrase, but they skillfully promulgated it, complete with framed and glossy photos, in support of their inspired environment of high-spirited excellence. I might add that their program properly credited the gentle Jedi professor. Here's another from AMD: "Performance is everything, but style counts." Whether you like that one is a sort of Myers-Briggs-like category detector. It's another delightful little ditty that lives just a step beyond traditional logic. But whoever said traditional notions of logic had to handle everything that matters and has power in life?

[2]  This is a universal resource locator (URL), the type of address used to find things on the World Wide Web. For further information about gaining access to this site, contact APICS.

3  For example, at this writing, the Yahoo search engine is a good starting point, at URL address http://www.yahoo.com. Yahoo provides access to quite a few other search engines, which, in turn, provide access to others. So this should be a good start.

4  You hadn't heard of the management approach called "Some Combination"? Guess you heard it here first.

5  Covey, S. R., *The Seven Habits of Highly Successful People*, Simon & Schuster, New York, 1989 (APICS item #03435).

6  Nietzsche said it once, as did my Uncle Charlie and Aunt Mildred. There's a lot that Nietzsche could have learned from my Uncle Charlie and Aunt Mildred.

7  At the Institute for International Research Theory of Constraints conference discussed in Chapter 4.

8  Hobbs, K., in the newsletter of the APICS Constraints Management SIG, *1996 APICS Atlantic Coast Symposium Proceedings*, and APICS' *The Production and Inventory Management Journal*.

9  Additional information about the synergies among TOC, Shingo methods, TPS, and lean manufacturing is included in the McMullen Associates web site at http://www.tbmcm.com.

10  For those not familiar with Detective Columbo, the character in the American television series and television movies, suffice to say he is a brilliant investigative mind with a disarming "Aw shucks" style of conversation and relationship that causes him to be underestimated by his opponents, to the opponents' chagrin.

11  A little mythology humor to lighten things up a bit. What's that? Right. Very little mythology humor.

12  *It's Not Luck* (APICS item #03291).

# Index

## A

ABC. *See* activity-based costing
Achterberg, Larry, 177
activity-based accounting systems, 24, 164
activity-based costing (ABC), 97, 118
activity-based management (ABM), 97
Adams, Scott, 16, 253, 265
Advanced Micro Devices (AMD), 278
agile manufacturing, 181, 213
airline computer system example, 92
Alcarez, Rachell, 186
Allied-Signal, 158
allocating costs, 223
allocation-based
    approaches, 102
    cost accounting, 24, 97
    product cost, 36, 101, 220, 223, 224
Amadas Industries, 214
Amelio, Gil, 166
American Society for Quality Control (ASQC), 10
Anderson, Melvin J., 145
Angst, David R., 127, 153
APICS, 10, 263
    web address, 204
APICS Constraints Management Specific
    Interest Group, 110, 148, 150, 164,
    205, 212, 213
APICS Constraints Management symposia,
    125, 148–158, 159, 205

APICS Constraints Management
    Symposium, 1995, 149–168, 229
APICS Constraints Management
    Symposium, 1996, 151–156, 189, 212,
    215, 229
APICS Constraints Management
    Symposium, 1997, 156–158, 213, 229,
    230
APICS Educational and Research
    Foundation, 164
APICS Educational Materials Catalog, 204
APICS international conferences, 159, 205
APICS Selected Readings in Constraints
    Management, 203
APICS Theory of Constraints Certifications,
    206
APICS Theory of Constraints Workshops,
    205
*Art of the Long View, The*, 93, 230
ASQC. *See* American Society for Quality
    Control
AT&T, 161
Atkinson, Thomas, 154
*Atlas Shrugged*, 251, 274
attention deficit disorder (ADD), 278
Attilla, 209
automotive glass industry, 163
Avery-Dennison Corporation, 125, 149,
    164, 189
Avraham Y. Goldratt Institute (AGI), 115.
    *See also* Goldratt Institute

# B

Bach, Richard, 141, 248, 258, 271, 272
Bal Seal Engineering, 128, 157, 215
Barr, Becky, 187
baseball/softball teams, 40
Baxter-Lessines, 162
beat a drum, 30
"being nice", 13
Benson, Herbert, 177
Berkeley, George, 251
Bescher, Bob, 130
best practices, 25, 51, 89, 111, 134, 136,
    143–145, 165, 232–234, 235
Bethlehem Steel Corporation, 116, 128, 158,
    266
*Beyond Good and Evil,* 258
Binney & Smith, 160
Bishop, John, 16, 118
Black, Joseph, 155
Block, Peter, 141, 209
Bloom, Bryan, 139
boards of directors, 136
    guidelines from and for, 135
Boeing Corporation, 156, 157, 261, 275
Bohnstengel, Bob, 187, 188
"Bookstore Effect", 17, 253
bookstores, 15–16, 18, 141
bottleneck, 30, 115
    as prime constraint, 108
breakthrough solutions, 45, 46, 50, 51, 56,
    88, 172
Brown, Beverly, 187
Buchwald, Steve, 164
buffer, 106, 132
buffer management, 43, 56, 63, 106–107,
    194, 197, 201
Burke, Edmund, 176
Burrus, Daniel, 16, 93
Bush, George, 10
business cycle, soft, 229
business plans, 40

# C

C/W Companies, 139
Camp, Glenn, 157

"can-do" attitude, 28, 45, 46, 64, 194, 228
*Candide,* 246
capacity
    constrained resource (CCR), 128
    exposing hidden, 44–45, 126–127
    finite, 105, 181, 182
        vs. inifinite capacity, 106, 108
    infinite, 182, 215
    requirements planning, 181
    capital charge, 98
Capra, Fritjof, 271
Carter, Jimmy, 10
cash flow, 35, 46
    high priority of growth, 137–138
Caspari, John, 156
categories of legitimate reservations (CLR),
    51, 59, 61, 71–72, 85, 143, 195
    causality existence reservation test, 71
    defined, 71–72
cause-and-effect, 12–13, 18, 22, 29, 32, 37,
    49, 51, 53, 57, 61, 62, 63, 64, 80, 92,
    94, 100, 101, 116, 118, 119, 135, 138,
    139, 153, 195, 223, 249
    thinking, and student essays, 187
Cavallaro, Harold, 201
CCR. *See* capacity constrained resource
*Celestine Prophecy, The,* 216, 271
Certified in Integrated Resource
    Management Accountants (CIRM),
    206
Certified in Production and Inventory
    Management (CPIM), 206
change agent, Eliyahu M. Goldratt as, 123
*Chaos,* 209, 275
Chesapeake Consulting, 157, 158, 259
chicken coop factory, 114, 147, 149
Chopra, Deepak, 177
Christensen, Roland, 16, 118
Christian Science, 177
Clara and Will example, 66
Clark, Ken, 16, 118, 208
Clemson University, 163
Clinton, Bill, 10
coherence, 139
Cole, Hugh, 157
collective unconscious, 119
Columbo, Detective, 31, 217–225

Committee on Research of the Institute of
  Management Accountants (CRIM),
  126
common sense, 18, 21, 51
  as common practice, 23, 125, 138
  constructing, 139
  economic, 135
  solutions, 46, 117
communications, 88, 172, 174
  improved by TOC, 140
  logic tree, 63
  simple, 177–178
compensation, 85
competition, 46
competitive positioning, 40
competitors, 8, 93
Computer Sciences Corporation, 144
conflict logic tree, 70
conflict-resolution diagram, 70
conservatives, 176
constraint, defined, 32
constraints, 59
  in a system, 32–33
  physical, 20, 32–33
  policy, 32–33
  thinking, 20
*Constraints Management Handbook*, 203
continuous improvement, 94, 224
Control Data Corporation (CDC), 114
controversy, regarding TOC, 9
core problems, 20, 44, 45, 56, 57, 173, 211,
  232
core process, 96
Corey, Ray, 118
*Corporate Comeback*, 253
correlation, 22
correlation approach, 18
cost accounting, 116, 217
cost-cutting, 100, 134
Cousins, Norman, 177
Covey, Stephen, 16, 87, 208, 209–212, 214,
  279
Covington, John, 141, 156, 158, 160, 203
Cox, James, 144, 151, 156, 203, 254, 275
CPM, 96, 109, 117
Crayola, 160
Creative Output, Inc., 114, 147

*Critical Chain*, 198, 199, 200
critical chain, 3, 4, 64, 89, 180, 183, 194, 196
  vs. critical path, 96–104
CRP, 152. *See also* capacity requirements
  planning
CRT. *See* current-reality logic trees
current-reality logic tree (CRT), 29, 45, 49,
  50, 55–61, 139, 140, 143, 153, 173,
  195, 198, 211, 240
  building, 59–60
  mutual fund diagram, 233
  ways to begin, 58–59


**D**

Danesi, Jr., Paul P., 126
Danos, Gerald, 126, 164
Darlington, John, 158
Davis, Fred, 16
decision support, 94, 154, 181, 182, 183
  software, 225
decision-making, 24, 28, 85, 86
  strategic, 224
delegation, 88, 172
Deloitte & Touche South Africa, 158
Delta Airlines, 144, 153
demand management, 235
Deming, W. Edwards, 208, 213
Department of Defense, 185
design of experiments (DOE), 202, 213
desirable effects (DE), 61, 62, 63, 64, 101
Detective Columbo. *See* Columbo, Detective
Detroit Public Schools, 186
Dettmer, William, 71, 142, 151, 158, 195,
  202, 213, 260
*Dianetics*, 271
dice game, 206
*Dilbert*, 16
*Dilbert Principle, The*, 253
dilemma logic tree, 70
direct labor
  as period cost, 98
Dixie Iron and Tool, 126, 164
"do-gooder", 11
Dossey, Larry, 177
Dresser Industries, 129, 157, 266
Drucker, Peter, 93, 118, 137, 211, 268, 269

drum, 105–106, 132
drum-buffer-rope (DBR), 25, 30, 43, 45, 56,
    63, 94, 106, 107, 129, 131, 132, 150,
    151, 156, 157, 167, 194, 196, 197, 201,
    206, 214, 236
    game, 207–208
due diligence, 9–10
dynamic buffering, 94, 108, 196
    factory scheduling, 43
Dynamic Manufacturing, 208

**E**

e-mail, 180
East Los Angeles, 186, 187
EC. *See* evaporating-cloud logic trees
economic value added, 24, 97, 98, 100, 102
ECT. *See* evaporating-cloud logic trees
ecumenism, 178–179
Eddy, Mary Baker, 273
Edelman, Phil and Hannah, 151
education, and TOC, 197–208. *See also* TOC
    for Education, Inc.
    conflict prevention, 186
    curricula, 186
    family relationships, 184
    role of teacher, 186
    study efficiency, 184
*Effective Executive, The*, 211
effectiveness, increased by TOC, 228
EG&G Corporation, 126, 165
Eglin Air Force Base, 184
80/20 rule, 38
elevating constraints, 43
Elwood City Forge, 165, 166
EMC Corporation, 144, 156
Emerson, Ralph Waldo, 245–246, 264, 265
"emotion of the inventor", 121
emotions, 61
    flow of, 48
employment stability, 31, 136
empowerment, 88, 172
    approaches, 141
enterprise resource planning (ERP), 94, 181,
    182, 203, 213, 215
environment, 179
Erhard, Werner, 141

ERP. *See* enterprise resource planning
establishing an objective, 75, 76
*Ethics*, 257
evaporating-cloud logic tree (ECT), 45, 49,
    50, 51, 65–74, 185, 198, 211, 240, 259,
    264
    diagrams, 198
    objective-needs-wants format, 66–68
    objective-requirements-prerequisites
        format, 66, 69
*Excellence Management*, 209
expediting, 57
expenditure control, 100
explicit assumptions, 51, 66
explicit use of thinking processes, 55
exploiting constraints, 43

**F**

factory scheduling, 43, 104–110
*Fifth Discipline, The*, 17, 87, 208
finite capacity, 105, 181, 182
    vs. infinite capacity, 106, 108
first-order principle, 53
fishbone diagrams, 50, 87
five-step focusing process, 29, 38–39, 42–43,
    44, 96–104, 150, 196, 235, 265
Flow International, 166, 276
focus, energy, 176
focusing, 118
    five-step process, 29, 38–39, 42–43, 44,
        96–104, 150, 196, 235, 265
    organizational, 87
    positive attitudes, 141–142
    processes, 166
Ford Motor Company, 19, 144, 160, 161,
    163
Forrester, J., 16
Foster, Warren, 151
*Fountainhead, The*, 274
Fox, Kevin, 184
Fox, Robert, 115, 151, 201, 254
Franklin, Ben, 86
FRT. *See* future-reality logic tree
future-reality logic tree (FRT), 26, 39, 40,
    45, 49, 50, 60, 61–62, 65, 143, 173,
    211

# G

Gadbois, Michel, 166
Gallagher, Neil, 128, 157, 160, 213
Garrison, Bill, 156, 164
General Electric, 114
General Motors, 19, 114, 116, 158, 160
generalists vs. specialists, 22
generally accepted accounting principles
        (GAAP), 223
genius, sharing process of, 122
Gerstner, Louis, 18
Ghemawat, Pankaj, 16
Gierman, Drew, 252
Glay, Cary, 189
Gleick, James, 209, 275
global measurements, 28–29, 34–36, 87,
        139, 196
global measures, 101
global performance, 182
goal, 135
        make money, 30–31
        of system, 28, 44
        of system, defined, 29
        operational, 35
Goal Systems International, 158, 259
*Goal, The,* 3, 4, 9, 28, 30, 31, 32, 35, 36, 102,
        105, 112, 115, 131, 142, 144, 156, 184,
        189, 197, 198, 199, 200, 201, 206, 214,
        215, 229, 238, 249, 254, 263
        conversation between Jonah and Alex,
        30, 32, 34, 102
goals, types of, 32
Goddard, Walter, 208
Goldratt Baseline TOC (GBT), 110
Goldratt, Eliyahu M., 2, 3, 19, 47, 58, 88, 98,
        104, 110, 111, 112–123, 131, 135, 146,
        147, 149, 151, 152, 156, 184, 200, 203,
        204, 215, 230, 247, 254, 258, 260, 261,
        263, 264
Goldratt Institute, 126, 147, 184. *See also*
        Avraham Y. Goldratt Institute (AGI)
*Goldratt's Theory of Constraints: A Systems
        Approach,* 202
"graduate student effect", 120
graduate students, 120–121
*Greatest Salesman in the World, The,* 249

Groulx, Marc, 157
Grove, Andy, 84
growth mentality, 134
growth strategies, 7
Grumman, 114
Gupta, Sanjeev, 151, 168
gut feelings, 21, 28

# H

Hamilton, Bruce, 215
Handerhan, Kevin, 165
Hannah's Do-Nut Shop, 19, 126, 151
Hansen Associates, 158
Hansen, Ray and Dawna, 156, 158
Hanson, Jesse, 188
hard sciences, 29, 47, 117
Harris Semiconductor, 127, 130, 152, 156,
        229
Harrison, Michael, 156
Harvard Business School, 17, 58, 172, 189
Hay, Edward, 208
Hayes, Robert, 16, 118, 208
*Haystack Syndrome, The,* 34, 35, 104, 116,
        128, 200, 202, 203, 254, 262, 263, 276
heart of the matter, 55–56
Henderson, Bruce, 16, 93, 118
Henry, Sr., Thomas R., 161
Herbie, 30, 105–106, 108, 158
Hermann Miller, 157
Hewlett-Packard, 19
*High Output Management,* 84
Hill, Napoleon, 141
Hinneburg, General Patricia A., 144, 155,
        275
Hobbs, Glenn, 157
Hobbs, Kermit, 214
Hofmans Forms Packaging, 162
Houle, Dale, 151, 184
Howmet Corporation, 131
Hoyle rules of Rummy, 72
human resources, 85
Hunsmanship, 209
Hunt, Ethan, 47–49, 51–52, 53–55, 59,
        64–65, 67, 69, 72, 73, 74, 76–80
        future-reality logic tree, 65
        prerequisites logic-tree diagram, 78

revised current-reality logic-tree
    diagram, 75
transition-tree diagram, 83
Hutchinson, Marcia, 187

**I**

i2 Technologies, 158
IBM, 18
idealist approaches, 142
idealistic vision, 175
idealists, 172
    activist, 175, 176
identifying obstacles, 75, 76
"if ... then" statements, 64
*Illusions*, 248, 258, 272
IMA. *See* Institute of Management
    Accountants
images, 52
immediate effects, 62
implementation, 89
    of TOC, 193
    process objectives, 195
    steps of TOC, 237
implicit assumptions, 51, 66
"impossible missions", 75
improvement curves, 39, 39–40
    applied to baseball batting, 41
    applying to anything, 40–41
*Inc.* magazine, 114
inertia, 29, 43, 44
    in thinking, 200
infinite capacity, 182, 215
information systems, 179–183
Ingersoll-Rand, 151
injections, 61, 62, 63, 68, 69, 81
inspirational motivator, Eliyahu M.
    Goldratt as, 122
instinct, 21, 80
Institute for International Research (IIR),
    159
Institute of Industrial Engineers (IIE), 166
Institute of Management Accountants
    (IMA), 10, 126, 276
    Report, 161–162
integration, 85
intermediate objective (IO), 76, 79, 82

International Society for Systems
    Improvement (ISSI), 162–163
Internet, 94, 204
intrinsic order, 28, 45–46, 54, 56, 59, 64, 117,
    143, 254
intuition, 49, 60, 73
    verbalizing, 21–22, 28, 61, 64, 65, 72, 140
intuitive styles of management, 3
intuitive use of thinking processes, 55
intuitives, 176
inventory, 34, 35
    reductions, 129–130, 153, 194, 214
investment, 34, 35, 98
Israni, Arjun, 127, 143, 150
*It's Not Luck*, 3, 4, 36, 116, 132, 135, 142,
    161, 162, 195, 197, 198, 199, 200, 232,
    238, 239, 240, 260
ITT, 116, 128, 157, 213
    AC Pump, 150
    Aerospace/Communications, 160

**J**

Japan, 17
    institutional shareowners, 137
Jenkins, Bob, 158
Johnson Controls, 19, 127, 152, 270
Jonah network, 127
Jonah process, 121
Jonathan Livingston Seagull, 272
Jung, Carl, 119
Juran, Joseph, 208
*Just In Time (JIT)*, 208
just-in-time (JIT), 118, 161, 181, 182, 203,
    213, 214–215
juvenile detention camp, 188

**K**

Kawasaki, Guy, 17, 252
Kendall, Gerald, 204
Kent Moore Cabinets, 159, 162, 228
Ketema A&E, 128, 129, 154, 266
Kibbey, David, 186
Kingman, Owen, 92, 153
Klein, Carl F., 127, 153
*Knots*, 178, 258, 278

knowledge base, 51
knowledge worker, 7–8, 227
Koziol, David, 167
Kroc, Ray, 264

**L**

labor as period operating expense, 225
labor as variable cost, 224–225
Laing, R. D., 177, 258
Laskey, Dan, 156
Lastname1-Lastname2, 86
layoffs, 230
    avoiding, 94
    eliminating need for, 134–138
lean management, 182
*Lean Manufacturing*, 208
lean manufacturing, 181
*Learning Organization*, 208
learning organization, 51
    process improvement, 212–213
LeFevre, Dennis, 131
legitimacy, 134
Levitt, Theodore, 16, 118
*Liberation Management*, 209
Lilienkamp, Dan, 162
Lincoln Electric, 119
local measurements, 29, 35–36, 87, 196
local measures, 139
local performance, 182
Locke, John, 251
logic, tight vs. loose, 62
logic trees, 2, 44, 47–88, 52, 212, 226, 235
    and case method, 189
    and information systems, 179–180
    and policy and strategy, 189
    applications of, 85
    as expression of scientific method, 117
    communication, 63
    conflict, 70
    conflict-resolution diagram, 70
    current-reality, 29, 45, 49, 50, 55–61, 139,
        140, 143, 153, 173, 195, 198, 211,
        240
        building, 59–60
        mutual fund diagram, 233
        ways to begin, 58–59

diagrams, 26, 27
dilemma, 70
evaporating-cloud, 45, 49, 50, 51,
        65–74, 185, 198, 211, 240, 259, 264
    diagrams, 198
    objective-needs-wants format, 66–68
    objective-requirements-prerequisites
        format, 66, 69
    format, 63
    four stages of, 52–55
    future-reality, 26, 39, 40, 45, 49, 50, 60,
        61–62, 143, 173, 211
        Ethan Hunt, 65
    generic vs. specific, 63
    learning process from do-ers, 74
    loose vs. tight, 260
    management processes, 7
    prerequisites, 45, 50, 51, 198, 210
        intermediate objectives (IO), 76, 77
    processes, 37, 38
    stage four, 54–55, 70, 73
    stage one, 52
    stage three, 53, 59, 70, 73
    stage two, 52
    study of rules, 72
    thinking processes, 43. *See also* thinking
        processes (TP)
    transition, 45, 50, 51, 80, 198
    trunk, 62, 65
    viewpoint, 63
    vision statement, 239
logistics, 104–110
logistics games, 206
loose vs. tight logic trees, 260
Lorrain, Marc A., 161
Low, Jim, 156, 158
low-hanging fruit, 95
loyalty, 134
*Loyalty Effect, The*, 135, 268
Loyola, Ignatius, 257
Lynch, Captain William, 155

**M**

Mackey, James, 202
making a difference, 12–13
"making the world a better place", 1, 13

management, 51
  as science, 15
  integrated system, 8
  integrating TOC, 208–217
  intuitive styles of, 3
  science, 2, 110
  science of TOC, 25
  skills, and TOC, 85, 88–89
  system, 62–65, 180
    elements of, 64
    elements of TOC, 26
    five-step introduction, 193–197
    foundations, 7
    introduction to, 205
    TOC as basis for, 23–24
  system, financial, 222
    building, 217–226
  three questions for, 37
managers
  and use of TOC, 8
*Managing for Results*, 118
Mandino, Og, 249
manufacturing
  continuous improvement, 43
  lean and agile, 43
  process improvement, 44–45
manufacturing resource planning, modern
    (MRP II), 94, 181, 182, 213, 215
marketing, 100
Marrus, Stephanie, 16, 93
materials requirements planning (MRP),
    181, 182, 203
Mavredes, William D., 160
McDonald's Corporation, 264
McKenna, Regis, 252
McKinney, Bryan L., 127, 153
McMullen Associates, 152, 158, 161, 204,
    259
Mealy, David, 167
meaningful coincidence, 216, 247, 252
measurements
  financial vs. non-financial, 37
  global, 28–29, 34–36
  local, 29, 35–36
  three fundamental TOC, 34
*Measurements for Effective Decision-Making*,
    202
meetings, better use of, 84–85

MEMC, 158
Mental Health Advocacy, 178
Mesone, Fred, 155
Michel, Robert, 16
Miles, R. H., 253
Millstein, Ira, 135
*Mission Impossible*, 47–50, 195, 227, 239
Mitchell, Robert, 157, 164, 213
Montgomery, Cynthia, 16, 93
Moon, Stacey, 154, 266
Moore, Kent, 159, 162
Moore, Richard, 150, 151, 156, 158
Moore, Shane, 186
morale, 134, 138
Mordoch, Avraham, 156
Morton International Automotive Safety
    Products, 130, 155
Morton Thiokol, 155
Mother Nature vs. Human Nature, 72
motivation, 20, 88
Moyers, Bill, 177
MRP, 152. *See* materials requirements
    planning (MRP)
mrp 1. *See* materials requirements planning
MRP II. *See* manufacturing resource
    planning, modern
Mueller, Robert K., 136
Murphy, Bob, 127, 152, 229
"Murphy's Law", 108
mutual fund industry, example, 231–232

**N**

Naisbitt, John, 16, 93
National Semiconductor, 19, 127, 143, 150,
    166
Navajo, 185
necessary conditions, 29, 36–37, 87, 135
needs demonstration entities, 81
negative branch processes, 65, 175, 176, 185,
    257, 264
net operating profit after tax (NOPAT), 98
net operating profit before tax (NOPBT), 98
net profit (NP), 98, 222
net sales price, 219
Newbold, Robert, 96, 204
Newtown, Kent, 158
Niceville High School, 186, 187

Nietzsche, Friedrich, 251, 258, 265, 273, 279
non-value-added activity, 45
Noreen, Eric, 202, 250, 261, 262, 263, 266
North Central Association Commission on
     Schools, 185
Northrup, Christiane, 177
Novotny, Donn, 156, 157
Nucor Steel, 119

**O**

Ohmae, Kenichi, 16, 118
Ohno, Dr., 208
Okaloosa County School System, 184
one-time changes in circumstances, 81
operating expense (OE), 34, 35, 98, 100,
     196, 222
operational effectiveness, 62, 63
organizations, classes of, 31
Overall, Karl, 125, 149
overhead allocation, 222–223, 224

**P**

"P&Q" exercise, 200
"P&Q" exercise, 128
P. L. Porter, 157, 166
Padmanabhan, Soundar, 158
Papp, Ken, 125, 149
Paresi, Barbara, 154
Pareto effect, 38
Pareto, Vilfredo, 255
Parr Instruments, 129, 130, 154, 266
Pasquarella, Marcia, 157
Peale, Norman Vincent, 141
Penrod, Bart, 155
PERT, 96, 109, 117
Peters, Tom, 16, 118, 144, 209
philosopher, Eliyahu M. Goldratt as, 122
physicist
    discovering simplicity, 24
    Elihayu M. Goldratt as, 122
    everyone as, 22
physics, 1, 263
    and performance improvement, 20
    of TOC, 15, 20, 23, 29
    TOC as, 119
Pickels, Dave, 157, 215, 266

Pirasteh, Reza, 129, 157, 266
Pirsig, John, 250
plan-do-check-act (PDCA), 213
planning, 89, 182
Plossl, George, 16
*poke yoke*, 209
policy development, 85
policy formulation, 51
Popcorn, Faith, 16
Porter, Michael, 16, 93, 118, 230
Posnack, Al, 156
Powell, Colin, 10
pragmatists, 175
Prahalad, Gary, 16
prerequisites logic tree (PRT), 45, 50, 51, 74,
    74–80, 198, 210
    intermediate objectives (IO), 76, 77
*Principle-Centered Leadership*, 208, 209, 210
Printronics, 157, 213
prioritizing, 211
problem solving, 51
process improvement, 43, 44–45, 85, 95–96,
    234–235
process management, 51
Procrustes, 228
Proctor & Gamble, 212
product development, 81, 100, 224
product mix, improving profitability of, 128
productive capacity, 229
productivity, 137
profitability, 56
    increased using TOC, 125–126
project management, 43, 51, 117
    critical chain, 25. See also critical chain
    using TOC, 89–96
    strategy, 89
*Project Management in the Fast Lane*, 96, 204
pros and cons, 65
protective capacity, 107–108, 229
PRT. *See* prerequisites logic trees
public domain, 148, 181, 183, 202

**Q**

QS-9000, 203
*Quality*, 208
quality improvements focused on TVA, 128
*Quest for Loyalty, The*, 268, 269

# R

*Race, The,* 201, 254
Rand, Ayn, 141, 246, 251, 274, 277
Rapone, Doug, 155
"ready-fire-aim" mode, 61
Redfield, James, 216
reengineering, 85, 100
*Reengineering the Manufacturing System,* 202
*Regaining Competitiveness,* 201, 206
Reichheld, Frederick, 16, 135, 268
Reid, Thomas, 251
*Relaxation Response, The,* 177
return-on-investment (ROI), 43, 98, 100,
　　109, 129, 131, 182, 222
revolutions
　　communication, 22
　　transportation, 22
revolutions in technology, 11
Richman, Dave, 167
Rickover, Admiral Hyman G., 112, 248, 261
　　Rickover Navy, 262
riddle, 243
Ridgeland Elementary School, 187
Ries, Al, 16, 93
Roadman, General Charles, 150, 152, 213
Robbins, Tony, 141
Roberts, Joe, 160
Robertson, Scott, 158, 202
Rockart, John, 16
Rockwell Automotive, 163
Rockwell Corporation, 19
Rogo, Alex, 131
Rohm & Haas, 163
root causes, 20, 44, 45, 56, 57, 87, 173, 200,
　　211
rope, 106, 133
Ruckel Middle School, 184, 188
*Running From Safety,* 272

# S

sales, 98
Samsonite Europe, 126, 130, 161
"saving the world", 10–11
SBDS. *See* schedule-based decision support
scale of importance, 12, 23, 35–36, 87, 96,
　　100–102, 196, 225

schedule-based decision support (SBDS),
　　35, 63, 64, 89, 94, 131
scheduling, 182, 183, 235, 236
　　breakthroughs, 114
　　technology, 104
Scheinkopf, Lisa, 149, 157
Schoenen, David, 128, 158
Schonberger, Richard, 208
Schragenheim, Eli, 156
Schumpeter, Joseph, 250
Schwarz, Peter, 16, 93, 230
science, 177
*Science and Health,* 273
scientific method, 2–3, 47, 116, 117, 247
*Securing the Future: Applying the Theory of
　　Constraints,* 204
Senge, Peter, 16, 87, 208
*Seven Habits of Highly Successful People,
　　The,* 209, 210, 279
Shapiro, Roy, 16, 118
Sheldrake, Rupert, 119, 272
Shingo Award conference, 215
Shingo, Shigeo, 128, 157, 182, 208, 215
Shoemaker, Larry, 151
Siegal, Bernie, 177
Sikorski, 114
Silva, Jose, 141
simple solutions, 5
simulations, computer-supported, 207
Sines, Sharon, 152, 229
single sufficiency demonstration entities, 81
Skinner, Wickham, 16, 118
small world, 174
Smith, Debra, 151, 202
Socratic method, 121
software for logic trees, 259
software vendors, 183
sounding board, 119
South Gross Pointe High School, 185
*Spectrum Management,* 151, 158
Spencer, Michael, 203
Spinoza, Benedict, 257
*Spiritual Exercises,* 257
Spodeck, Gil, 156
Srikanth, M. L., 151, 201, 202
stakeholders vs. shareholders, 99
Stanley Furniture, 160
statistical process control (SPC), 202, 213

Stein, Robert, 151, 202, 213
Stentz, Will, 185
Stewardship, 209
Stillahn, Brad, 167
strategic
    analysis, 89
    decision-making, 224
    effectiveness, 62, 63
    planning, 43, 51, 85, 92, 118
    positioning, 40, 89, 194
strategist, mind of, 8
structured thinking processes, 87
subconscious minds, 119
subordinate, as part of five-step focusing
    process, 43
*Success* magazine, 144
Suerken, Kathy, 157, 184–185, 188, 278
Sumter Cabinet Company, 163
Sundararajan, Sekar, 167
supply chain, 109, 129–130, 156, 181, 182,
    183, 193, 194, 205
*Synchronous Manufacturing*, 201
synchronous manufacturing, 106, 133, 151,
    157
synchronous remanufacturing, 154
synergy, 212
system
    as defined by TOC, 28
    defining, 29
    goal, defined, 29
    identifying constraints, 43
    information, 179–183, 183
    management, 180
    thinking, 28, 139–140

**T**

Taguchi Loss Function, 213
Tandem Computer, 269
teamwork, 88
Texas Instruments, 19, 126
*Theory of Constraints (TOC), The*, 202
*Theory of Constraints and Its Implications for
    Management Accounting*, 250
"think globally and act locally", 12
thinking, 61
    and health, 176–178
    inertia, 200

progress of, 55
tools, 26, 116, 180
types of, 48–49
thinking processes (TP), 21, 29, 32, 47–88,
    115, 116, 132, 150, 152, 153, 172,
    175–176, 179, 185, 194, 196, 202, 205,
    226, 235, 246, 250
and management skills, 88–89
applications of, 2
benefits of, 51–52
defined, 2
revealing intrinsic order, 46
used with five-step focusing, 43
thoughts, flow of, 48
three-ring binders, 238–239
throughput, increasing, 45
throughput value added (TVA), 31, 35, 36,
    93, 96, 100, 126, 127, 131, 135, 164,
    194, 196, 201, 211, 217, 220, 222
accounting management, 25, 222
defined, 31, 97
financial management, 1, 7, 31
financial management system, 24–25, 89,
    94, 97–104, 222
financial management system,
    diagrammed, 100
sample TVA report, 222
Thru-Put Technologies, 158, 164, 166
time
    as prime constraint, 41, 84–85, 256
    buffers, 106, 108, 109, 201
    using well, 45
*Timeless Healing*, 177
Tipper Tie, 166
*TOC and Its Implications for Management
    Accounting*, 202
TOC Center, 259
TOC for Education, Inc., 157, 184–188
    web address, 189
Toffler, Alvin and Heidi, 16, 93
*Tom Sawyer*, 245
tool selection, 49–50, 55
Toro, Anthony J., 151
Torrington Company, 151
*Total Quality Management*, 208
Total Quality Management (TQM), 50, 87,
    128, 155, 160, 161, 165, 202
    processes, guided by TOC, 213

totally variable costs (TVC), 97, 98, 220, 221

*Tough Fabric*, 203

Toyota production system (TPS), 182, 208, 214, 215

TP. *See* thinking processes

TQM. *See* Total Quality Management

trade unions, 137–138

Transcendentalist, 246

transition logic tree (TT), 45, 50, 51, 80–84, 198

Treybig, Jim, 269

TRIZ, 87–89, 261

Trout, David, 16, 93

true economic value added, 102

TT. *See* transition logic trees

Turner, Ted, 10

TVA, 253. *See* throughput value added

TVA financial management system, 194, 205, 225–226, 236

TVA throughput world, 36, 37, 135, 137, 200

TVA-I-OE, 230, 232, 269

chart, 222

TVC. *See* totally variable costs

**U**

U.S. Air Force, 158

U.S. Air Force Logistics Command, 144, 155, 275

U.S. Air Force Medical Service, 152, 213, 229

U.S. Department of Commerce Center for Excellence, 158

U.S. Naval Academy, 189

U.S. Navy, 158

Ulowetz, Kirsten, 158

Umble, Michael, 201

undesirable effects (UDEs), 56, 57, 59, 198

United Air Lines, 154

United Electric Controls, 215

United Technologies Corporation, 130

**V**

validity vs. truth, 21

Valmont Industries, 167

Van Langenhoven, Thomas, 144

vendor management, 85

verbalizing intuition, 49, 61, 64, 65, 72, 140

verbalizing thoughts, 48

vision statement trees, 239

vision statements, 62

Vollum, Robert, 156

Vornlocker, Robert, 150, 276

**W**

Walker, Ben, 185

*War and Peace*, 277

Wayman, Willard A., 155

wealth-producing capacity, 137

Webb, Jr., Larry E., 160

Weil, Andrew, 177

Weisz Graphics, 163

well-being, 21, 62, 177

West Tape and Label, 163, 167

Western Textile Products, 162

*What Is This Thing Called the Theory of Constraints and How Should It Be Implemented?*, 200, 264

Wheatley, Margaret, 141, 209, 275

Wheelwright, Steven, 16, 118, 208

Whirlpool Corporation, 167

Wigand Corporation, 126

Wilmot, Jeanine, 158

Wilson, Ken, 154

win-win solutions, 22–23

"wisdom to know the difference", 171–172, 173

Woeppel, Mark, 156

work centers

non-constraint, 107, 108, 133

synchronization of, 165

work in process, 106, 168

*World as Will and Idea, The*, 264

*World Class Manufacturing*, 208

world class manufacturing (WCM), 214

World Wide Web (WWW), 95, 180, 278

*Wow! Management*, 209

"wows", 246. *See also* Peters, Tom

**X**

Xerox El Segundo, 168

## Y

Yip, George, 16, 93

## Z

*Zarathustra,* 273
*Zen and the Art of Motorcycle Maintenance,*
    250
Zycon Corporation, 126, 127, 128, 129, 130,
    151, 163